Synthesis Lectures on Welding Engineering

Series Editors

Menachem Kimchi, State Library of Ohio, Columbus, USA

David H. Phillips, Columbus, USA

This series publishes short books on fundamentals, principles, and applications on a variety of Welding Engineering topics including spot welding, solid-state welding, a guide to navigating welding codes, computational modelling of welds and welding processes, laser welding, arc welding automation and control, and micro-joining. Initial topics have been chosen because they are either not well represented by the current market of published material, and/or similar publications are outdated and do not represent the latest welding advancements, knowledge, and technology.

Menachem Kimchi · David H. Phillips

Resistance Spot Welding

Fundamentals and Applications for the
Automotive Industry

Second Edition

 Springer

Menachem Kimchi
Welding Engineering Program
The Ohio State University
Columbus, OH, USA

David H. Phillips
Welding Engineering Program
The Ohio State University
Columbus, OH, USA

Synthesis Lectures on Welding Engineering
ISBN 978-3-031-25782-7   ISBN 978-3-031-25783-4   (eBook)
https://doi.org/10.1007/978-3-031-25783-4

This Springer imprint is published by the registered company Springer Nature Switzerland AG
The registered company address is: Gewerbestrasse 11, 6330 Cham, Switzerland

Preface

This book was written by two professors in the Welding Engineering program at Ohio State University in response to the rapidly increasing challenge for Resistance Spot Welding the many families of advanced materials in the modern automotive industry. The aggressive implementation of these materials is driven by the ongoing need to develop vehicles that are lighter weight (for better fuel efficiency) and safer to drive. Both goals rely heavily on stronger, lighter weight materials. In particular, modern steels known as "advanced high-strength steels" are rapidly becoming integral to automotive construction. This family of steels offers an excellent combination of high strength and ductility. The high strength of these materials allows for the use of thinner sheets and structures to reduce weight, as well as to provide improved crash protection in certain parts of the car. Aluminum alloy applications are increasing as well, primarily due to their excellent strength-to-weight ratio. Even advanced aluminum alloys that previously were used primarily for aerospace applications are being considered.

To successfully implement these materials, in many cases they will have to be spot welded, a welding process that the automotive industry relies heavily on to keep vehicle costs relatively low. Whereas resistance spot welding of plain carbon steels is well established and well understood, advanced high-strength steels and aluminum alloys present many new challenges to the welding engineer. To date, there is no textbook that covers these specific topics, and the coverage of the resistance spot welding process in general is limited. The intent of this book is to fill in some of those gaps.

Columbus, USA   Menachem Kimchi
                                                                              David H. Phillips

Contents

About the Author

Menachem Kimchi is an Associate Professor of Practice in the Department of Material Science & Engineering, Welding Engineering Program at The Ohio State University. Menachem has over 30 years of experience in the automotive industry, and previously he was a principal research engineer and technology leader for EWI. Menachem has been involved extensively with research and development projects for the Automotive and Steel industries and published over 100 technical papers in the areas of Resistance and Solid State Welding processes of advanced materials. He has also served on AWS (American Welding Society) Resistance Welding and Automotive committees and chaired the IIW (International Institute of Welding) Sub-commission on Resistance Welding.

Dr. David H. Phillips is a Professor of Practice in the Welding Engineering program at The Ohio State University. After completing his Master's degree in Welding Engineering in 1986, he spent approximately 20 years in industry, with most of his experience in the aerospace and automotive sectors. David then earned his Ph.D. in Welding Engineering and began teaching at Ohio State in 2008 where he now teaches a variety of welding processes and weld metallurgy courses from sophomore to graduate level. David is a licensed Professional Engineer (P.E.) in Welding Engineering, an International Welding Engineer (I.W.E.), and was a Certified Welding Inspector (C.W.I.) for 12 years. He is also the author of a recently published textbook titled *Welding Engineering: An Introduction.*

Introduction to Resistance Spot Welding

Resistance Spot Welding (RSW) is one of a family of industrial welding processes that produce heat required for welding through what is known as joule ($J = I^2Rt$) heating. Much in the way a piece of wire heats up when current is passed through it, a resistance spot weld forms due to the resistance heating that occurs when current passes through the parts (or sheets) being welded. For this reason, metals with higher resistivity such as steel are more easily resistance welded than those that are better conductors of electricity such as aluminum.

There are many resistance welding processes, but the most common and the subject of this book is RSW (Fig. 1.1). This process relies on opposing copper-based water-cooled electrodes that serve two functions-provide clamping force and passing electrical current through the two sheets being welded. If the sheets are steel, the resistance to the flow of current through the sheets will be much higher than the copper electrodes, so the steel will get hot while the electrodes remain relatively cool. The result is the formation of an elliptical shaped weld between the sheets known as a nugget.

The three important spot welding process parameters are weld current, weld time, and electrode force. Important spot welding terminology includes squeeze time, weld time, hold time, and off time. Squeeze time is the amount of time needed to develop the required electrode force, weld time is the time in which current flows, hold time is the time to allow weld solidification while the force is still applied, and off time is the time needed for the electrodes to move to the next weld. Weld cycles may be relatively simple and utilize just the aforementioned times, but they can often be much more complex (Fig. 1.2). Modern machine controllers offer the ability to customize and precisely control each of the parameters indicated on the figure. RSW works extremely well for welding the relatively thin sheets that are so common in automotive manufacturing. In addition, incredibly fast welding times combined with the self-clamping nature of the process make

© The Author(s), under exclusive license to Springer Nature Switzerland AG 2023   1
M. Kimchi and D. H. Phillips, *Resistance Spot Welding*, Synthesis Lectures
on Welding Engineering, https://doi.org/10.1007/978-3-031-25783-4_1

Fig. 1.1 Resistance spot welding

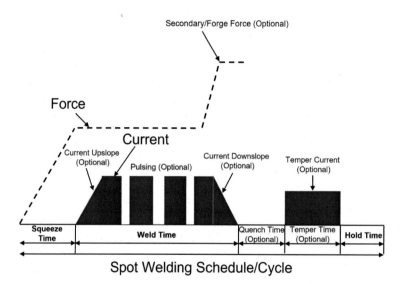

Fig. 1.2 RSW cycles always include squeeze, weld, and hold times, but they are often more complex

it an ideal process for high production robotic welding on which the automotive industry relies heavily.

Resistance Spot Welding Process Physics and Fundamentals

2

2.1 Mechanism of Heat Generation and Nugget Growth

The concepts of resistance and resistivity both play a critical role in the RSW process. Prior to discussing the fundamentals of heat generation during a spot weld, it is important to first clarify the difference between resistance and resistivity. Resistivity is a physical property of a material such as metal that refers to how strongly the metal opposes (or resists) the flow of electrical current. The symbol most used for resistivity is ρ. Resistance incorporates the physical property of resistivity, but also includes the geometry of the material (conductor) that is passing current. Much like the flow of water through a pipe, smaller diameters and longer lengths create greater resistance. Therefore, the formula for resistance is $R = \rho l/A$, where $\rho =$ the resistivity, $l =$ the length of the conductor, and $A =$ the area of the conductor.

Heat generation associated with the RSW process is more complicated than just the simple heating of a conductor associated with the passage of electrical current. Figure 2.1 again shows the standard RSW arrangement of Fig. 1.1, but includes a plot of resistance across the electrodes that reveals a very important characteristic critical to most resistance welding processes-the contact resistance between the sheets being welded. As indicated on the figure, the highest resistance to the flow of current is where the sheets come into contact with each other. This fact allows a weld nugget to begin forming and grow exactly where it is needed-between the sheets. A typical approach to determining various resistances is shown in Fig. 2.2.

To better understand the heat generation mechanism, the current path from one electrode to the other can be compared to an electrical circuit that contains seven "resistors" in series. In Fig. 2.1, "resistors" 1 and 7 represent the bulk resistance of the copper electrodes, "resistors" 2 and 6 represent the contact resistance between the electrodes and the sheets, "resistors" 3 and 5 represent the bulk resistance of the sheets, and "resistor 4"

© The Author(s), under exclusive license to Springer Nature Switzerland AG 2023
M. Kimchi and D. H. Phillips, *Resistance Spot Welding*, Synthesis Lectures on Welding Engineering, https://doi.org/10.1007/978-3-031-25783-4_2

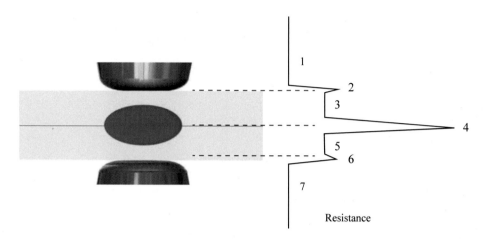

Fig. 2.1 Relative resistance across a resistance spot weld prior to nugget formation

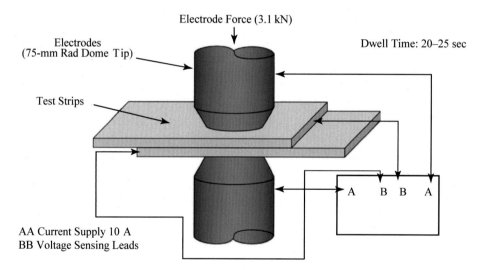

Fig. 2.2 Typical approach to measuring resistances associated with RSW [*Source* WorldAutoSteel]

represents the contact resistance between the sheets mentioned above. It is important to point out that while the figure shows the location of a successful weld (called a nugget), the plot of relative resistance describes the situation at the very beginning of the weld, and prior to nugget formation. As described later, these resistances change rapidly during the short time it takes to produce a single weld, a concept referred to as dynamic resistance.

Another factor that works strongly in the favor of resistance welding of steel is that as steel is heated, its resistivity relative to copper increases dramatically, as shown in

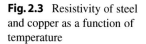

Fig. 2.3 Resistivity of steel and copper as a function of temperature

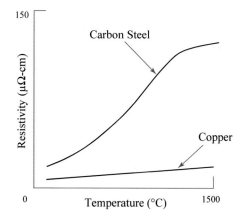

Fig. 2.3. So, the initial contact resistance effectively heats the surrounding area, which is in turn heated more rapidly by the flow of current because its resistivity is higher due to the higher temperature. As a result, heating and weld nugget formation can occur quite rapidly. A typical spot weld time for steel of about 1 mm thickness is approximately 1/5 of a second. The welding current required for RSW is much higher than for arc welding, with a typical current range for the RSW process of 8–15 kA for steel.

2.2 Effect of Electrode Force on Contact Resistance

The previous discussion emphasizes the important of contact resistance between the sheets being welded when producing a spot weld. As shown in Fig. 2.4, the amount of contact resistance depends on both the surface condition of the sheets being welded, and the applied electrode force (or pressure). This figure explains part of the reason for the high contact resistance between the sheets being welded, and the subsequent effect of electrode pressure during welding. All metals have some surface roughness that will limit the contact area when the two sheets are brought together. As a result, when current is passed through the sheets, the electrons will be forced to flow through the narrow regions where the surface asperities touch. This creates localized extremes in current density that causes an increase in resistance. When electrode force is applied, surface asperities are collapsed, and the contact resistance becomes less. This means that with RSW, higher forces will generally result in less heating due to reduced contact resistance. In addition to surface roughness, other conditions that increase contact resistance include oxides, rust, scale, grease/oil, and paint. Higher electrode forces will have the same effect of lowering the contact resistance by breaking up oxides and pushing out other surface contaminants.

Therefore, it seems logical that to keep contact resistances high, the lowest electrode forces possible would be the most ideal. However, Fig. 2.5, a plot of contact resistance versus electrode force, indicates why this is not the case. As mentioned, an important

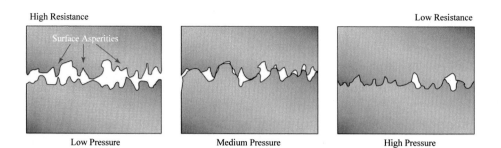

Fig. 2.4 The effect of electrode force on contact resistance

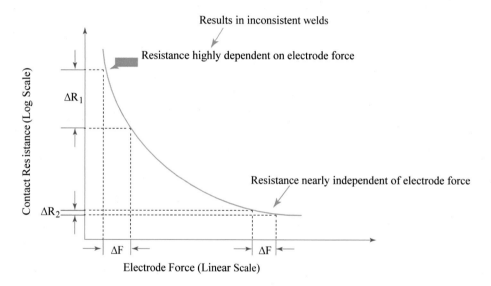

Fig. 2.5 The relationship between contact resistance and electrode force

relationship associated with the RSW processes is the relationship between contact resistance between the two sheets being welded, and the applied electrode force. The figure shows that at extremely low electrode forces, small changes in force will produce large changes in contact resistance. Since some spot welding machines utilize less-than-precise pneumatic cylinders to create force, there can be some small variations in force from part-to-part resulting in large changes in contact resistance. However, an even more important issue is if there is any variation in the surfaces of the sheets being welded (as is common in production), these variations will result in inconsistent heating when very low forces are being used. The result will be a narrow process window for producing good welds, and generally inconsistent weld quality. Higher forces are therefore needed to achieve consistent welds. However, if the forces are too high, part indentation and possibly electrode wear can be a big problem. These are general relationships only; the amount of force will depend on many factors including the strength of the material being welded. But in summary, when choosing an electrode force, it is important to find a balance that addresses the discussed concerns and produces the widest process window possible.

2.3   Current Range and Lobe Curves

Visual examination is often an important and effective method for verifying the quality of welds such as arc welds. However, with most resistance welding processes, visible examination is not possible due to the "hidden" weld location since the welds are located between the sheets or parts being welded. As a result, maintaining weld quality with processes such as RSW is highly dependent on other approaches such as the use of current range (Fig. 2.6) and Lobe (Fig. 2.7) curves. The current range curve is produced experimentally, and the lobe curve is generated from multiple current range curves using different weld times. Both the current range and lobe curves reveal ranges of weld current/time that will produce acceptable welds in production, as well as provide for a means of quality control through the monitoring of these weld parameters.

The current range curve shows the relationship between nugget (weld) size and weld time, while the lobe curves establish a process window for producing acceptable nuggets. Since nugget size (diameter) directly affects the strength of the weld, it is an important measure of whether the weld is acceptable. A general rule of thumb for the minimum

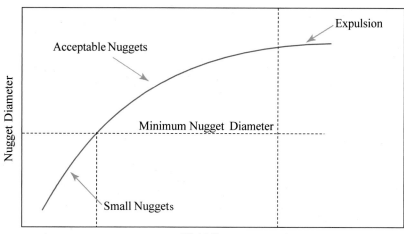

Fig. 2.6 Current range curve

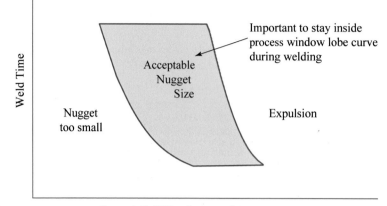

Fig. 2.7 A lobe curve for the RSW process

size weld nugget diameter is $4\sqrt{t}$ (where t = sheet thickness in millimeters), but the minimum size can vary depending on design requirements. On both the current range and lobe curves, the maximum current is defined by excessive expulsion, which refers to tiny drops of molten metal that are forcibly expelled from between the sheets due to the weld size becoming too large (due to excessive heating), or insufficient electrode force. The current range curves can show that a spot weld can be produced in extremely short times which would seem to be the goal in automotive production. However, excessively high currents and short times will produce a narrower process window. This trend is indicated on the lobe curve, which begins to become narrower at the shorter times on the bottom portion of the curve.

As mentioned, the lobe curve is developed experimentally by producing a series of current range curves, as shown in Fig. 2.8. To produce the first nugget diameter versus current curves, a series of spot welds is made at increasing levels of weld current while maintaining the same weld time. A common and simple form of testing referred to as a peel test (discussed later in the chapter on quality control) is used to pull welded coupons apart and measure the nugget diameter. The current level producing the minimum nugget size is identified as a point on the curve, as well as a point indicating maximum current where excessive expulsion is observed. These two points are then plotted below on a plot of weld time versus weld current, using the weld time used to produce the curve. This produces the first two points on the lobe curve. Subsequent weld trials to produce nugget diameter versus current curves conducted at various times are then plotted the same way to complete the maximum and minimum current lines of the lobe curve. The lines indicating maximum and minimum times are not determined experimentally and are somewhat arbitrary.

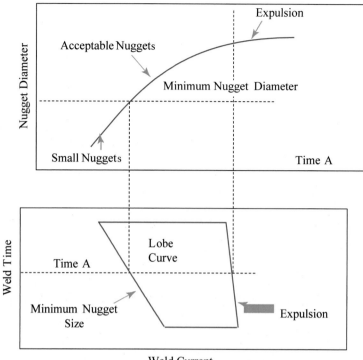

Fig. 2.8 Experimental approach to developing a lobe curve

2.4   Expulsion

There are two mechanisms associated with expulsion. The first mechanism occurs when the weld simply gets too large for the pressure ring created by the electrodes. This can happen when the weld time and/or weld current is excessive. The second expulsion mechanism occurs when there is insufficient electrode force. In this case, the pressure created from the expanding weld nugget (due to volumetric expansion when a solid metal becomes liquid) exceeds the electrode pressure forcing the sheets apart allowing molten metal to spray out.

Expulsion is a common occurrence in automotive production. A reasonable amount of expulsion is often considered acceptable since it provides evidence that a weld is being made while also producing welds of decent strength. However, the problem with relying on expulsion is that weld strengths are likely to be less consistent than when there is no expulsion. This is often compensated for by producing additional welds. Severe expulsion should always be avoided since it produces significant indention and poor-quality welds.

So, although some amount of expulsion continues to be acceptable in many production facilities, there is a general trend toward eliminating expulsion as an accepted practice for the reasons just discussed. Eliminating expulsion may be more achievable today with modern machines equipped with controllers that provide for much better control and feedback of the process parameters.

2.5   Determination of Spot Welding Parameters for Production

The development of current range and lobe curves usually begins with a review of pertinent welding standards (such as AWS C1.1 discussed later in this book), which provide tables of recommended parameters including time, force, and approximate current. But there are so many variables (Table 2.1) associated with spot welding, these tables generally should not be used solely for determining production parameters. In particular, whereas weld time and force included on the tables may be acceptable for production, an acceptable current range must be determined experimentally. And whereas the resulting curve provides a range of weld current that will produce acceptable welds in laboratory conditions, decisions must still be made regarding the selection of a production weld current. A common approach regarding the selection of a production weld current is to choose a current from the lobe curve that is just below the expulsion level.

The above approach is probably the simplest approach to determining production spot welding parameters. But in some cases, the entire set of parameters may be determined experimentally by developing various curves that also include the effect of force. In this way, electrode force can be optimized to develop the widest process window that will produce an acceptable weld. This produces a set of variables that will be the most "robust" to variations in production conditions, a common situation in automobile manufacturing.

Another method which has historically been used as a starting point for developing weld parameters is known as the "law of thermal similarity." It can be useful when a change of sheet thickness is made to a well-established weld schedule. This law states that weld time is proportional to the square of the thickness of the sheets being welded. For example, if the thickness of the sheets being welded is doubled, the weld time should be increased by four times. In addition, current and electrode size should be doubled. This approach is based mainly on differences in bulk heating due to changes in sheet thickness and doesn't consider contact resistances between the sheets. Therefore, it is generally not very accurate and should be used with caution.

Table 2.1 The many variables of spot welding

Material	Process	Machine	Maintenance
Material properties • Electrical resistivity • Thermal conductivity • Heat capacity • Density • Thermal expansion coefficient • Strength	*Weld specifications* • Strength • Diameter • Expulsion • Indentation • Location • Flange width	*Electrical* • Current balance • Power factor • AC, DC, and IQR • Loop magnetism • Supply voltage	*Electrical* • Actual loop resistance • Actual loop area
Microstructural factors • Cracking • Grain growth • Hardness • Heat-affected zone • Phase changes	*Weld parameters/Variables* • Current • Electrode force • Weld time • Squeeze time • Electrode selection • Shunting • Electrode wear	*Mechanical* • Air and Servo pressure • Gun design • Force balance • Cooling system • Misalignment	*Mechanical* • Mechanical tool wear • Lubrication • Cooling water flow • Misalignment
Geometric factors • Thickness • Thickness ratio • Sheet stack-ups • Flatness • Metal fit			*Die practice* • Metal fit • Costing damage • Flange condition
Surface factors • Roughness • Oil • Oxide • Coating type • Paint • Misc. contamination			

[*Source* WorldAutoSteel]

2.6   Dynamic Resistance

As discussed previously, the resistance across a spot weld during the time the weld is being made is not constant but changes rapidly, and therefore is considered "dynamic." By monitoring the resistance during a single spot weld, a plot of resistance versus weld time can be generated. The result is known as a dynamic resistance curve since it reveals how much resistance changes over the course of a weld. A typical dynamic resistance curve for carbon steel is shown in Fig. 2.9 and includes additional information (in the circles) to indicate what is occurring at any region of the curve. The dynamic resistance plot aids in the understanding of resistance welding fundamentals, as well as providing a method for weld quality control.

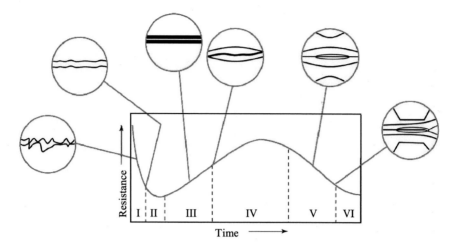

Fig. 2.9 A typical dynamic resistance curve for spot welding steel

As indicated on the left side of the figure, resistance starts out relatively high, drops, and then begins to rise again. This portion of the curve reveals the transition from contact resistance to bulk resistance. Initially, contact resistance is high as surface oxides and rough surfaces create greater resistance to the flow of current. With enough pressure and heat, the oxides are shattered, and the surface asperity peaks collapse, effectively lowering the contact resistance. As discussed previously, the contact resistance heating rapidly heats the surrounding steel, which dramatically increases its resistivity, and bulk resistance of the sheets begins to dominate. The bulk resistance heating soon becomes large enough to begin forming a molten nugget. As the nugget grows, the current path becomes larger which reduces the current density, which causes the resistance to plateau and then drop. A further drop in resistance may occur if there is indentation since the length of the current path is reduced. If there is expulsion, severe indentation may occur dropping the resistance even more. As discussed later in Chap. 8, these curves may be used to monitor and control weld quality.

2.7   Heat Balance

A successful spot weld requires that the nugget be centered between the sheets being welded. This is usually relatively easy to achieve when the two sheets being welded are of similar thickness and made of similar metals. However, there are many occasions when this is not the case creating a heat balance problem causing the nugget to shift in one direction or the other. A simple spot welding heat balance problem and possible solution is shown in Fig. 2.10. When welding a thicker sheet (top sheet in the figure) to a thinner

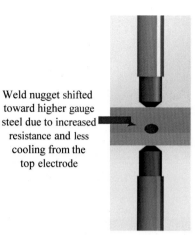

Weld nugget shifted
toward higher gauge
steel due to increased
resistance and less
cooling from the
top electrode

Larger diameter top electrode
reduces current density and
increases cooling so the
nugget shifts to the center

Fig. 2.10   Heat balance problem on the left, solution on the right

sheet, the nugget will tend to shift toward the thicker sheet. This is mainly because the current path is longer in the thick sheet so resistance is higher ($R = \rho l/A$, and in this case, l = thickness of the sheets). Depending on how thick the sheets are, the upward shift may also be influenced by less cooling from the top electrode since it is farther away from the nugget. One approach to balancing this situation is shown on the right side of the figure in which a bigger diameter top electrode is used. This will reduce current density through the top sheet (by increasing area A in the resistance formula), which decreases the heating in the top sheet causing the nugget to shift toward the lower sheet and become more centered. A simple technique to determine electrode sizes in a heat balance situation is shown in Fig. 2.11. By drawing two straight lines starting from the outer diameter of the large electrode that cross where the two sheets meet, the diameter of the smaller electrode can be determined. A similar heat balance problem will occur if the top sheet has a greater resistivity than the bottom sheet. Another possible solution to this example is to choose a lower electrode with greater resistivity than the upper electrode. In the automotive industry, heat balance problems often get even more complex than the example shown (such as with multiple sheet stack-ups), and these will be discussed in Chap. 7.

Fig. 2.11 A simple approach to determining electrode face diameters to improve heat balance when welding sheets of two different thicknesses

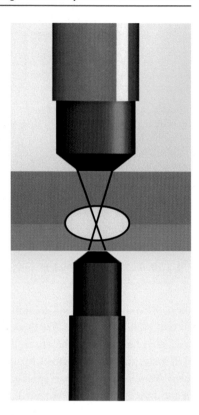

Resistance Spot Welding Machines, Electrodes, and Tooling

3

3.1   Machine Types

RSW machines are available in a wide variety of configurations, from relatively simple rocker arm machines which rely on foot power to provide the force, to much more sophisticated robotic systems with electric servo guns. Power supplies range from single phase AC machines to three phase medium frequency DC (MFDC) inverter power supplies. The choice of machine/power supply is of course a function of many factors such as production requirements, type and thickness of material being welded, economics, etc. But as indicated on Fig. 3.1, most RSW machines have four common features: (1) the step-down transformer which converts high incoming voltages to low voltages (and high currents) in the secondary which is where the welding electrodes are located; (2) a mechanism for delivering the required electrode force; (3) a water supply and required plumbing for cooling both the transformer and the electrodes; and (4) a weld controller which has the basic function of receiving operator inputs and running the machine.

The most common type of RSW machines in the modern automotive assembly plant are the robotic MFDC-powered machines utilizing pneumatic, or more commonly now, electric servo guns. While welding with these machines will be the focus of this book, it is helpful to first do a quick review of all of the RSW machine types.

The simplest machine configuration, the rocker arm-type, is shown in Fig. 3.2. These machines offer long horns (the arms that hold the electrodes), and therefore long throat depths, which provide flexibility for welding a large range of part sizes. The top horn moves in an arc while the lower horn is stationary. Electrode force may be generated mechanically by manually depressing a foot pedal, or more commonly using a pneumatic

M. Kimchi and D. H. Phillips, *Resistance Spot Welding*, Synthesis Lectures on Welding Engineering, https://doi.org/10.1007/978-3-031-25783-4_3

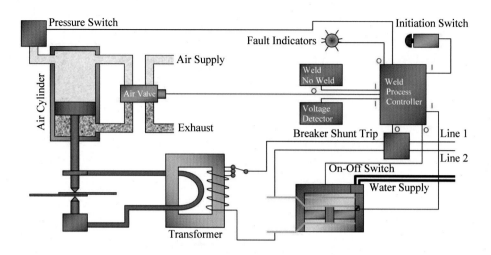

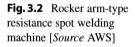

Fig. 3.1 Basic RSW machine set-up [*Source* EWI]

Fig. 3.2 Rocker arm-type resistance spot welding machine [*Source* AWS]

cylinder as is the case with the machine in the figure. Because of the arcing movement, electrode alignment relative to the parts can be a problem. The long horns can also be prone to flexing which can lead to electrode skidding. While these machines are

Fig. 3.3 Press-type resistance
spot welding machine

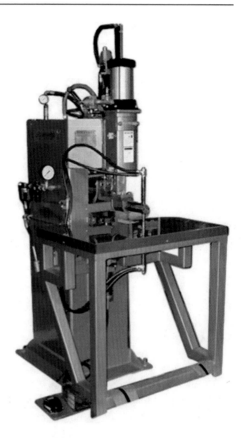

inexpensive and versatile, they are mainly applicable to low production, job shop type
applications, and/or for building prototypes.

Figure 3.3 shows another type of machine known as a press-type machine. These are
very rigidly constructed machines that utilize a moving top head that travels in a straight
line, as opposed to the arcing movement of a rocker arm type machine. To accomplish
this movement, the secondary of the transformer must also include a flexible current
shunt. Because of their rigid construction and linear motion, they are not susceptible to
the problems associated with rocker arm-type machines. They provide for very precise
electrode alignment, which also makes them ideal for another type of resistance welding
known as resistance projection welding. Electrode force is usually provided by pneumatic
cylinders, but when extremely high forces are needed, hydraulic cylinders may be used.
Press-type machines can weld a wide range of materials and thicknesses, but again, are
not conducive to high-speed production.

Fig. 3.4 MFDC robotic spot welder [*Source* T.J. Snow]

As mentioned, the focus of this book is with RSW associated with modern robotic MFDC powered machines. A typical machine is shown in Fig. 3.4, and a complete system is shown in Fig. 3.5. As will be discussed in the next section on power supplies, MFDC (inverter) power supply designs provide for an incredible reduction in transformer size and weight. This is extremely important when spot welding in a high production environment since it means the transformer can be mounted at the end of the robot arm, greatly increasing electrical efficiency. The gun that holds the electrodes generates clamping force either pneumatically, or using electric servo motors. The electric servo guns (Fig. 3.6) are becoming much more common in the automotive industry as they offer many improvements over pneumatic guns such as faster welding times, more precise control of electrode clamping force, and the ability to vary force during a single weld. However, pneumatic guns may still be preferred in some applications. A comparison of pneumatic and electric servo guns is included in Chap. 9.

3.2   Power Supplies

This section will briefly summarize the power supplies available for RSW. Compared to arc welding power supplies, RSW power supplies (transformers) produce much higher secondary currents and utilize lower voltages. Because spot welding currents can be 10,000 amps or more, the electrical efficiency, or power factor, is often an important

Fig. 3.5 Robotic spot welding system [*Source* Kuka]

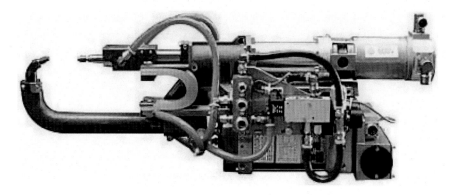

Fig. 3.6 Electric servo spot welding gun [*Source* ARO]

consideration. Each power supply type offers advantages and disadvantages, but only the MFDC power supplies can be made small enough for robotic applications that are so important to the automotive industry. All power supplies may be either single or three phase, but in industrial applications they are almost always three-phase.

The simplest power supply is an AC power supply. These are also the least expensive since their transformers are relatively simple. However, they do not offer precise control of the welding process, and their electrical efficiency is poor due to the high reactance in the secondary. Reactance is only associated with alternating current and acts similar to resistance in a DC circuit. Reactance is also a function of the frequency of the AC-the higher the frequency, the higher the reactance, and the lower the electrical efficiency or power factor. In summary, for this reason, AC machines are known to be the least efficient of all the RSW power supplies.

Another machine known as a frequency converter offers improved electrical efficiency over AC machines through the reduction of the AC frequency in the secondary. Whereas a standard AC machine has a frequency of 60 Hz in the secondary, the frequency converter offers frequencies typically less than 20 Hz. For the reasons mentioned, these machines offer improved efficiencies due to the reduced reactance in the secondary.

DC power supplies offer the best electrical efficiency and the most precise control of the spot welding process. In particular, as mentioned, the MFDC machines have become the power supplies of choice in automotive production plants. These power supply designs are based on the principle that by increasing the electrical frequency in the primary side of the transformer, the transformer becomes much more efficient, and components (such as the core) can be made much smaller. The "M" in MFDC refers to the "medium" frequencies of 1,000–2,000 Hz (or higher), which are generated in the primary side of the transformer via a sophisticated controller and a network of solid-state switches. The tremendous reduction in size and weight of the transformer allows it to be mounted on the end of the robot arm, creating a short path for the welding current to the electrodes. In addition to these advantages, DC welding current provides for much more control and weld consistency than AC current.

Another less common family of resistance welding power supplies that can deliver extremely high current pulses for very short times are known as capacitive discharge (CD) machines. These are mostly used in projection welding applications, but may be beneficial in some spot welding applications, such as when there is a need to minimize the marking of the surface that is in contact with the electrodes.

3.3   Electrode Materials

The choice of electrode material for a given spot welding application is an important consideration. Electrodes must be able to conduct current to the part, mechanically constrain the part, conduct heat from the part, and provide for long life. They must be able to sustain high loads at elevated temperatures, while maintaining adequate thermal and electrical conductivity. The choice of electrode alloy for a given application is often dictated by the need to minimize electrode wear while accomplishing the required functions. When electrodes wear (see Chap. 9), they typically begin to "mushroom," or grow larger in diameter. Electrode wear is accelerated when there is an alloying reaction between the electrode and the part, a common problem when welding aluminum and coated steels. The problem becomes even worse if the electrodes are not cooled properly by the internal water-cooling system. As the electrode diameter increases, the current density decreases, resulting in a decrease in the size of the weld. Since the strength of a spot weld is directly related to the size (diameter) of the weld nugget, electrode wear can be a big problem since it can result in welded parts with strength below what is required.

A range of copper-based or refractory-based electrode materials are used depending on the application. The Resistance Welding Manufacturers Association (RWMA) sorts electrode materials (for all resistance welding processes) into three groups: A, B, and C. Group A includes electrode materials most commonly used for RSW. They are copper rich with small amounts of alloying elements added, such as zirconium, cadmium, chromium, and beryllium. Group B contains refractory metals such as tungsten and molybdenum, and refractory metal powder metal composites with copper, and Group C are copper electrodes dispersion strengthened with aluminum oxide. Within each of the groups, they are further categorized by a class number. The general rule of thumb is as the class number goes up, the electrode strength goes up, but the electrical conductivity goes down. When electrical conductivity goes down, the electrode will get hotter more easily resulting in premature electrode wear. The choice of electrode material involves many factors, but generally higher strength electrodes will be selected when higher strength materials are being welded. To keep electrodes from getting too hot, it is also important that the electrical and thermal conductivities of the electrodes are much higher than those of the material being welded. Electrodes may be forged or cast, but they are most commonly forged.

The most used spot welding electrodes are the Class 1 and 2 Group A electrodes. The first three classes of Group A electrodes are shown on Table 3.1, which reveals the relationship between increasing strength and decreasing electrical/thermal conductivity with increasing class numbers. The International Annealed Copper Standard (IACS) column on the table refers to a copper standard regarding electrical conductivity to which the electrodes are compared, where pure Cu has an IACS number of 100%. Class 1 electrodes

Table 3.1 Group A Electrodes (Cold-Worked)

Electrode Diameter		Conductivity – IACS (%)			Rockwell Hardness (HRB)			Ultimate Tensile Strength (ksi)		
Size Range		Class			Class			Class		
(in.)	(mm)	1	2	3	1	2	3	1	2	3
Up to 1	Up to 25	80	75	45	65	75	90	60	65	95
Over 1 to 2	Over 25 to 51	80	75	45	60	70	90	55	59	92

offer the highest electrical and thermal conductivity, and are commonly used for welding high conductivity materials such as aluminum, magnesium, brass, and bronze. Class 2 electrodes are considered general purpose since they can be used for a wide range of materials, including carbon, low alloy, and stainless steels. These electrodes contain additions of mostly chromium, but sometimes contain zirconium, or a combination of both. Class 3 electrodes are not as common, but are the best choice when higher strength electrodes are needed (at the expense of lower conductivity). The Class 20 aluminum oxide dispersion-strengthened electrodes of Group C offer similar conductivity and strength to the Class 2 electrodes but maintain their strength to higher temperatures. Therefore, these more expensive electrodes may be selected for welding coated steels which are known to cause rapid electrode wear. Figure 3.7 illustrates the trade-off between conductivity and relative strength (determined by hardness) as a function of temperature for these common electrodes, and includes the Copper Association designation numbers. In summary of this section, the spot welding electrode alloy choice is driven by a balance of all of these factors and depends highly on the material being welded. Subsequent chapters of this book will include electrode alloy recommendations for each specific family of materials to be discussed.

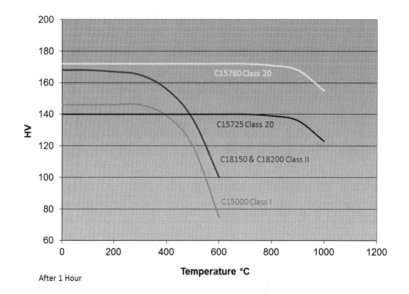

Fig. 3.7 Effect of temperature on hardness (HV) for common Class A electrodes

3.4 Electrode Geometries

In addition to electrode alloy type, a choice must be made regarding its shape. Figure 3.8 shows the standard spot welding electrode shapes per RWMA, but many different versions of these shapes and other shapes are used. A proper electrode geometry for a given application is one that optimizes the electrical-thermal-mechanical performance of an electrode. Other factors may include considerations such as accessibility to the part and how much surface marking of the part is acceptable. For example, the Type "D" offset shape is designed to place a weld close to a flange, whereas Type "C" or "D" designs may be chosen to minimize part marking. The diameter of the electrode contact area is also a consideration. Too small an area for the material and thicknesses being welded will produce undersized welds with insufficient strength, while too large an area may lead to inconsistent weld growth characteristics.

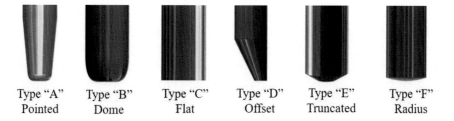

| Type "A" | Type "B" | Type "C" | Type "D" | Type "E" | Type "F" |
| Pointed | Dome | Flat | Offset | Truncated | Radius |

Fig. 3.8 Softening curve for common electrodes [*Source* Luvata]

Electrodes may be single piece electrodes, but quite often they are made up of two pieces consisting of an adapter shank and a cap tip. This approach greatly reduces electrode replacement costs. The adapter shanks may be designed for either male or female cap tips. A design incorporating a female cap tip is shown in Fig. 3.9. This figure also reveals the internal cavity of the electrode which provides for water cooling which is so important to long electrode life. Figure 3.10 shows actual examples of conventional and experimental electrode designs.

Fig. 3.9 Two-piece RSW electrode with female-type connection

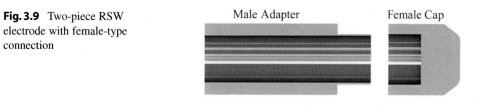

Male Adapter   Female Cap

Fig. 3.10 Examples of spot welding electrode cap designs with a male-type connection

   The electrodes must then be connected to the electrode holders on the machine. The most common approach is a tapered, press-fit connection. The RWMA Resistance Welding Manual provides guidance on these connections and categorizes the various taper options. Figure 3.11 reveals the details associated with the various tapers, while Fig. 3.12 combines recommended tapers with electrode and electrode face diameters as a function of electrode force. This obviously reflects the fact that higher electrode forces will require large diameter electrodes and electrode faces, and the size of the electrode dictates the taper to be used.

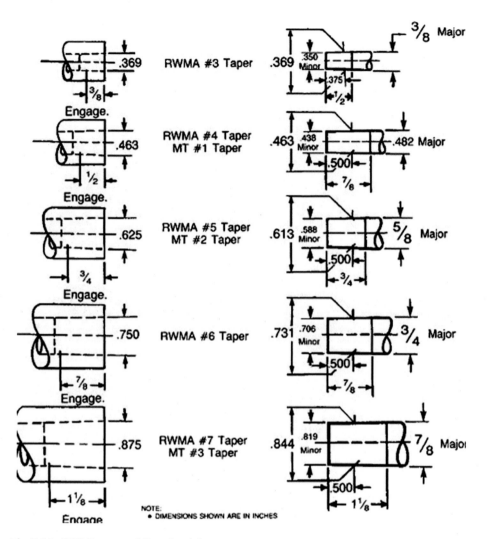

Fig. 3.11 RWMA spot welding electrode tapers

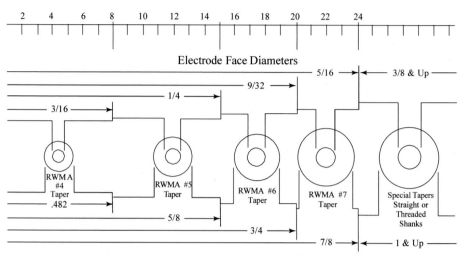

Fig. 3.12 RWMA taper, electrode diameter, and electrode face diameter recommendations for spot welding electrodes as a function of electrode force

Resistance Spot Welding of Modern Steels

4

4.1 Introduction to Automotive Steels

As discussed in the Preface, the automotive industry's motivation to build safer, lighter-weight, and more fuel-efficient vehicles is driving new alloy development, in particular the development of new steels. Whereas the general public's impression of steels might not be "high-tech," the reality is that steels are unique in that they offer such a wide variety of strengthening mechanisms, excellent formability, and low cost. Steel makers today continue to develop new automotive steels, with the general goal of producing higher strengths while maintaining ductility.

The wide range of steels available to automotive manufactures is best described on a plot of strength versus elongation (or ductility). A version of this plot, which is informally referred to as the "banana chart" because of its shape, is shown in Fig. 4.1. The plot clearly reveals the trade-off between elongation and strength, and illustrates the wide range of steel alloy families available. Such a wide variety of strength/ductility combinations provides automotive manufactures with the opportunity to customize their choice of steel for each component on the vehicle. Whereas some components require excellent ductility because of the extensive forming operations that may be required, others may need higher strength for reducing weight (higher strength steels can be made thinner) and/or improving crash protection. And some aspects of crash protection benefit from steels with greater ductility that can absorb impact better. Such a wide variety of steels makes life more complicated for the Welding Engineer, who must address spot welding of many of these steels, the subject of this chapter.

M. Kimchi and D. H. Phillips, *Resistance Spot Welding*, Synthesis Lectures on Welding Engineering, https://doi.org/10.1007/978-3-031-25783-4_4

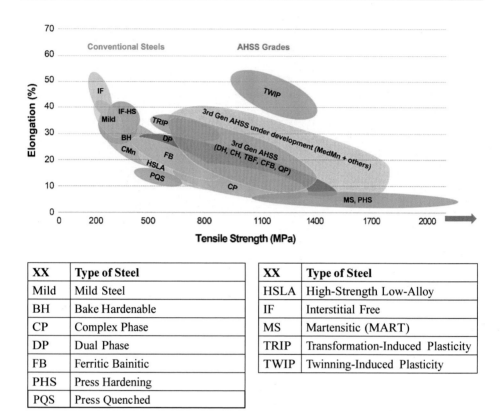

XX	Type of Steel
Mild	Mild Steel
BH	Bake Hardenable
CP	Complex Phase
DP	Dual Phase
FB	Ferritic Bainitic
PHS	Press Hardening
PQS	Press Quenched

XX	Type of Steel
HSLA	High-Strength Low-Alloy
IF	Interstitial Free
MS	Martensitic (MART)
TRIP	Transformation-Induced Plasticity
TWIP	Twinning-Induced Plasticity

Fig. 4.1 Automotive steels as a function of % elongation and tensile strength [Courtesy of WorldAuto-Steel]

4.2   Conventional Steels

Conventional steels, which consist primarily of mild (or plain carbon), interstitial free (IF), and high-strength low-alloy (HSLA) steels are the simplest in terms of alloy content and are generally the easiest to spot weld. As the figure shows, they also offer the highest ductility, but at the expense of low strength (as compared to the advanced high-strength steels). Carbon contents are generally in the range of 0.01–0.45 weight percent but are sometimes higher. As with all steels, they contain the very important strengthening element manganese, but in small amounts, typically greater than 0.3 weight percent and less than 2 weight percent. Other elements present in even smaller amounts in mild and interstitial free steels include silicon, sulfur, and phosphorous. HSLA steels offer greater strength primarily due to the addition of elements such as vanadium, titanium, niobium, and copper. These elements, in combination with special processing techniques, produce greater strengths through grain refinement of the ferritic microstructure. Ferritic-Bainitic

(FB) steels are like HSLA steels in that they rely on grain refinement for strengthening, but contain an additional harder phase known as bainite. As a result, FB steels offer both improved strength over conventional HSLA steels, as well as better ductility when comparing FB and HSLA alloys of similar strength. IF steels contain the lowest amount of carbon (~0.01%) and manganese (~1.0%) and are used when extremely high ductility is needed for forming purposes. These are also sometimes called "deep draw" steels.

4.3   Welding of Conventional Uncoated Steels

As mentioned, conventional uncoated steels are generally the easiest metal alloys to spot weld. Modern versions of these steels are even easier to weld than their predecessors due to vast improvements in steel making practices. Historically, spot welding of conventional steels sometimes produced a form of cracking known as solidification cracking. Susceptibility to this form of cracking increased when impurities of sulfur and phosphorous in the steel became too high. This is no longer a significant issue since steel producers have learned how to keep these impurity levels very low.

The primary issue with spot welding conventional steels is the concern of forming hard martensite, which results in brittle nuggets that will not perform well in a peel test. Although conventional steels generally don't form martensite easily, the cooling rates associated with the spot welding process can be extremely high (>1,000 °C/s). In fact, such cooling rates can be more than 100 times the cooling rates associated with arc welding. And the thinner the sheets being welded, the faster the cooling rates. This is because the water-cooled, highly conductive copper electrodes play an even more significant role on the cooling rate as the sheets get thinner. Such high cooling rates can produce martensite in steels that are normally not considered hardenable, such as conventional steels. When martensite does form, its hardness is directly related to how much carbon the steel contains, so high carbon-containing conventional steels will be the most susceptible to forming hard martensite.

A common way to assess the likelihood of forming martensite in conventional steels is to utilize carbon equivalency (CE) formulas. These involve making a simple calculation based on carbon content and other elements in the steel. The greater the CE, the greater the concern for forming hard martensite. There are many versions of this formula, but a well-known version is the International Institute of Welding (IIW) version:

$$CE = \%C + \%Mn/6 + \%(Cr + \%Mo + \%V)/5 + \%(Si + Ni + Cu)/15.$$

These formulas should be used primarily for comparative purposes since in spot welding, the cooling rate is probably the most dominant factor in terms of the likelihood of martensite formation.

But in general, a CE (using this formula) of approximately 0.3–0.35 is the range in which martensite formation begins to be a concern.

Table 4.1 Adapted from AWS C1.1:2019 recommended practices for resistance welding. Spot welding parameters for bare and zinc-coated steel < 350 MPa ultimate tensile strength*

Metal thickness (mm)	Electrode face diameter (mm)	Net electrode force (kN)	Coated weld time (mS)	Bare weld time (mS)	Approximate welding current coated (amps)	Approximate welding current bare (amps)
0.5	4.8	1.8	170	120	11,000	8,000
1.0	6.4	3.1	230	170	14,000	12,000
2.0	7.9	6.2	420	300	20,000	17,000

* *Note* This table of a random sampling of typical parameters serves as a rough guide and starting point only. Also, refer to the C1.1 standard for much more complete and comprehensive details

Table 4.1 adapted from the RWMA Resistance Welding manual provides some guidance for welding low carbon conventional steel (SAW 1008-1010) sheets of varying thickness. When developing a set of spot welding parameters, this type of chart can be considered a good starting point. It is important to note that the table emphasizes that welding current is approximated because of potential wide variations in resistances primarily associated with different heats of steel which often have variable surface conditions. Therefore, to determine the required current for each application current ranges should be developed experimentally. Other rules of thumb exist for spot welding conventional steels, such as welding time (cycles) = 10t(in mm) + 2, and welding force (kN) = 2.45t, where t = the thickness of the sheets, but usually the optimum parameters must be developed for each application since there are so many variables to consider. It's also important to note that with all metals, the greater the alloying content, the higher the resistance to the flow of electrical current. This means that alloys with smaller amounts of alloying elements will generally require greater welding current and/or time. For example, while the table recommends about 14,000 amps for an approximately 2-mm thick sheet of 1010 steel, higher amperages (and possibly longer times) would be needed for an IF steel of the same thickness since IF steels have the least amount of alloying elements. This disparity is depicted in Fig. 4.2, which shows the approximate location of Lobe curves for two different steels.

4.4 Advanced High-Strength Steels

The subject of modern advanced high-strength steels requires a more thorough review of the processing, metallurgy, and microstructures prior to discussing approaches to spot welding. Compared to conventional steels, advanced high-strength steels contain higher alloy additions with complex microstructures consisting of multiple phases. They all rely on some combination of processing at high temperatures (thermomechanical processing)

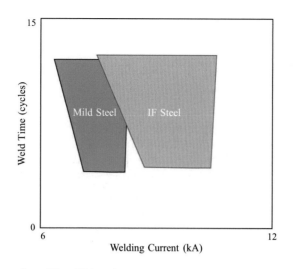

Fig. 4.2 Lobe curves for mild and IF steels

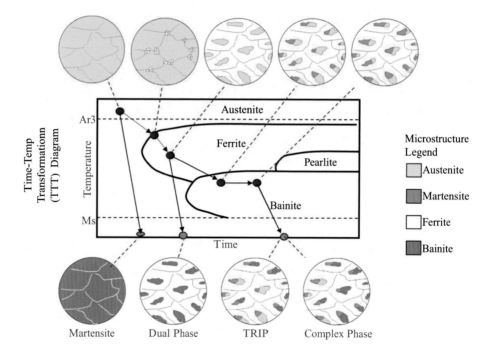

Fig. 4.3 Advanced high-strength steels rely on various processing techniques that begin with heating into the austenite phase field [*Source* WorldAutoSteel]

in the austenite phase field followed by controlled cooling (Fig. 4.3). The ability to combine alloying with thermomechanical processing techniques allows for the customization of microstructures, and therefore customization of properties to achieve the best combination of strength and ductility. They also may be further strengthened significantly by strain hardening. For proprietary reasons, actual chemical compositions of these steels are not always published. As indicated previously on Fig. 4.1, the numerous families of advanced high-strength steels offer strengths ranging from approximately 500 to over 2,000 MPa, and % elongations as high as 60% to less than 10%. The steels that will be covered in this chapter are the transformation-induced plasticity (TRIP), dual phase (DP), complex phase (CP), martensitic (MS), twinning-induced plasticity (TWIP), and press hardening (PHS) steels. These steels are typically classified (or graded) by the steel type and tensile strength (in MPa), and not by their composition. For example, DP 500 is a dual phase steel with a tensile strength of 500 MPa. Current research and development in steel manufacturing is focusing on a new family of steels referred to as third generation advanced high-strength steels. The primary goal with the next generation of steels is to increase ductility even further while maintaining high strengths.

4.4.1 Transformation-Induced Plasticity (TRIP) Steels

TRIP steels consist of a complex microstructure of mostly soft ferrite, with smaller amounts of harder phases bainite and martensite, and a minimum of 5% by volume retained austenite (Fig. 4.4). The retained austenite is achieved primarily by adding higher amounts of carbon, which is an austenite stabilizer. The processing steps include an isothermal hold to produce the bainite. The most beneficial and important feature of TRIP steel is its excellent work hardening through strain-induced transformation of the retained austenite to martensite, such as during forming, or in a crash. Greater amounts of strain produce even more transformation to martensite, and the level of strain at which transformation begins is a function of carbon content. This means TRIP alloys with low carbon contents can be used more for complex shapes that allow the forming operation to strengthen the part through work hardening. By adding higher amounts of carbon, retained austenite remains in the part, and will only be transformed to martensite under extreme levels of strain, such as during crash deformation. So by varying carbon content, TRIP alloys may offer both excellent ductility and strengthening from work hardening. This strengthening is beneficial for both the forming of the part, and during a crash where additional hardening occurs from the high amounts of strain.

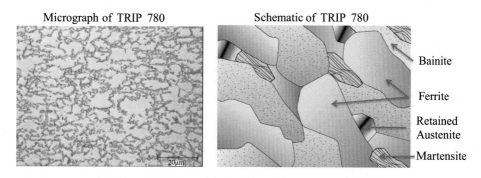

Micrograph of TRIP 780   Schematic of TRIP 780

Bainite

Ferrite

Retained
Austenite

Martensite

Fig. 4.4 TRIP steel microstructure [*Source* WorldAutoSteel]

4.4.2 Dual Phase (DP) Steels

The microstructure of DP steels is a mixture of two phases-hard martensite and soft ferrite (Fig. 4.5). This is achieved by controlled cooling from austenite temperatures to initially allow for some transformation to ferrite, which is then followed by a quench to transform the remaining austenite to martensite. The soft, continuous ferrite gives these steels excellent formability, and during forming, most of the strain occurs in the ferrite which provides for good work hardening. The strength/ductility ratio can be increased by increasing the ratio of martensite-to-ferrite. In the very high-strength DP steels such as DP980, the amount of martensite may exceed 50%, and the ferrite is no longer continuous. DP steels achieve additional hardening known as bake hardening. The forming operation produces dislocations which promote the formation of carbides during the subsequent curing operation in paint bake ovens. TRIP steels can also be further strengthened by bake hardening.

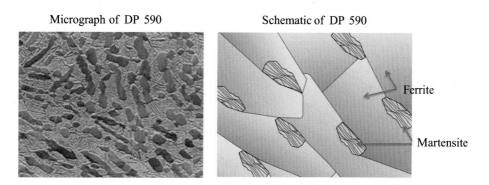

Micrograph of DP 590   Schematic of DP 590

Ferrite

Martensite

Fig. 4.5 DP steel microstructure [*Source* WorldAutoSteel]

Micrograph of CP 800/1000   Schematic of CP 800/1000

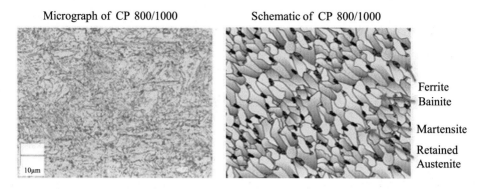

Ferrite
Bainite

Martensite

Retained
Austenite

Fig. 4.6 The fine, complex microstructure of CP steel [*Source* WorldAutoSteel]

4.4.3 Complex Phase (CP) Steels

CP steels offer very high strengths due to an extremely fine grain structure consisting of a ferrite/bainite matrix, with small amounts of martensite, pearlite, and retained austenite (Fig. 4.6). The extremely fine grain sizes are achieved by the precipitation of elements such as titanium and niobium, which act to delay recrystallization and pin grain boundaries during processing. CP steels offer greater strength than TRIP steels and many DP steels, but at reduced ductility.

4.4.4 Martensitic (MS) Steels

MS steels consist of mostly martensite with small amounts of ferrite and/or bainite (Fig. 4.7) produced by rapid quenching from austenite temperatures. Elements such as silicon, boron, chromium, nickel, manganese, molybdenum, and vanadium in certain combinations are added to increase hardenability. They are among the highest strength (approaching 1,700 MPa) of all the advanced high-strength steels, but with limited ductility. As with martensite in conventional steels, the strength of MS steels is directly related to carbon content. A tempering heat treatment can be utilized after quenching to recover some ductility to create the best balance of strength and ductility.

4.4.5 Twinning-Induced Plasticity (TWIP) Steels

The microstructure of TWIP steels consists entirely of twinned austenite (Fig. 4.8). Austenite in these steels is stabilized down to room temperature via additions of very high amounts (17–24%) of the austenite stabilizing element manganese. These steels can be deformed (strained) through a process known as mechanical twinning, which differs

Microstructure of MS 950/1200   Schematic of MS 950/1200

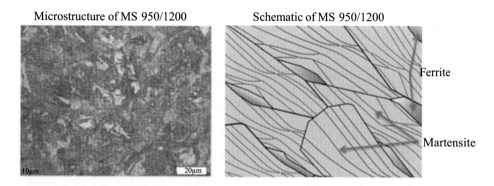

Fig. 4.7 MS steel microstructure consisting primarily of martensite [*Source* WorldAutoSteel]

from standard individual dislocation movement during straining. This deformation through twinning offers extremely high strain-hardening rates due to the rapid increase in the fraction of deformation twins during straining, which act much like grain boundaries to effectively block dislocation movement. The combination of a soft austenitic grain structure prior to forming and the extremely high strain hardening due to mechanical twining produces the widest range of ductility and strength of any advanced high-strength steel. The more ductile versions of TWIP steels can produce ductility levels as high as 60% elongation (exceeding even IF conventional steels), while the higher strength versions can approach tensile strengths of 1,300 MPa.

Micrograph of TWIP Steel
(as annealed)   Schematic of TWIP

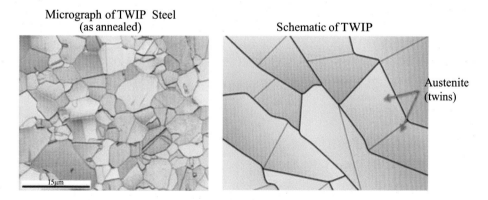

Fig. 4.8 TWIP microstructure of twinned austenite [*Source* WorldAutoSteel]

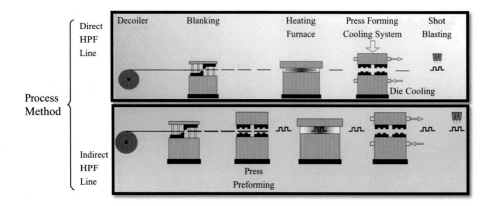

Fig. 4.9 The two approaches to hot press forming of boron steels [*Source* WorldAutoSteel]

4.4.6   Press Hardening Steels (PHS)

PHS steels are similar to MS steels in that their extremely high strengths (up to 2,000 MPa) come from a microstructure of martensite. These steels are produced using a process known has Hot Press Forming (Fig. 4.9). When utilizing the direct Hot Press Forming method shown in the figure, most or all of the forming is done while the steel is in the austenitic temperature range (Fig. 4.10). With the indirect method, some of the forming is done at room temperature, but final forming is conducted at austenite temperatures. Yield strengths at these temperatures are very low making it easy to form complex shapes. In both cases, the cool dies in the final forming operation quench the steel to form martensite, so in effect, the heat treating and forming operations are combined into a single operation. In some cases, cooling rates may be tailored to produce different properties in different locations of the part. In most cases, very small amounts of boron (0.002–0.005 weight %) are added to these steels to improve hardenability by making it easier to form martensite per a given cooling rate. This is the reason these steels are sometimes referred to as boron steels or hot-stamped boron steels.

4.4.7   Ferritic-Bainitic (FB) Steels

FB steels represent another family of steels, like DP steels, that rely on two phases, but in this case the two phases are ferrite and bainite. These microstructures (Fig. 4.11) are generally finer than the DP steels, and offer an even greater ability to customize and optimize the balance between excellent ductility/formability and good strength. They perform well under fatigue loading conditions and are therefore often utilized in components that incur dynamic loads.

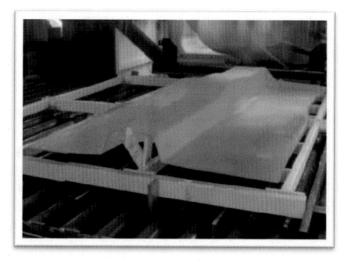

Fig. 4.10 PHS steel during a hot press forming operation [*Source* WorldAutoSteel]

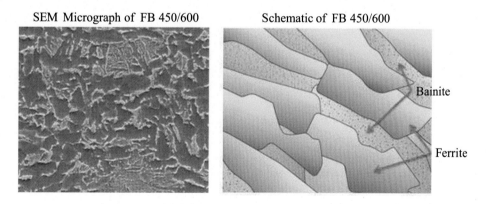

Fig. 4.11 Microstructure of FB steels consist of a mix of ferrite and bainite [*Source* WorldAutoSteel]

4.5 Welding of Advanced High-Strength Steels

4.5.1 Spot Welding Parameters

Because advanced high-strength steels are more highly alloyed than conventional steels, their resistivity is higher meaning less spot welding current will usually be needed to produce a weld nugget. Figure 4.12 shows the relative shift in current levels that can be expected when transitioning from welding conventional steels to advanced high-strength

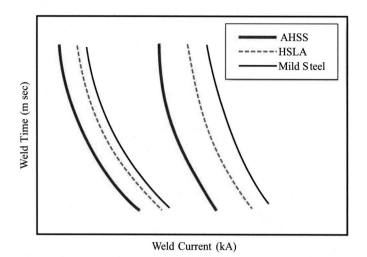

Weld Time (m sec)

Weld Current (kA)

Fig. 4.12 Typical lobe curves for advanced high-strength steels reveal the need for less current than with conventional steels [*Source* WorldAutoSteel]

steels. And because they are much higher in yield strength, higher forces are needed to produce proper contact between the sheets. Electrode forces will need to be 20% higher (or more), and larger diameter electrodes may be required. Class 2 electrodes are almost always used. Larger welds may be required as well, so $5\sqrt{t}$ may be a better minimum nugget size guide, as compared to $4\sqrt{t}$ for conventional steels.

The figure also reveals another issue when welding these steels-the process window (current range) tends to be narrower than with conventional steels. This means it is more difficult to create a minimum-sized nugget without expulsion. To increase the current range, and therefore the process operating window, higher forces and/or higher times are needed. These general relationships between current range and forces/times are shown in Figs. 4.13 and 4.14 which are plots from actual experimental data on advanced high-strength steels. AWS C1.1-"Recommended Practices for Resistance Welding"-provides spot welding guidelines for advanced high-strength steels based on the strength of the steel being welded. Table 4.2 shows C1.1 recommended parameters for steels with tensile strengths between 350 and 700 MPa, while Table 4.3 is for steels with strengths greater than 700 MPa.

Another industrial guideline to spot welding parameters (provided by WorldAutoSteel, and originally developed by General Motors) is as follows:

- t = Metal Thickness (inch)
- y_s = Yield Strength of Steel (KSI)
- d = Contact Tip Diameter (inch) = $\sqrt{1.65t - 0.007}$
- A = Contact Tip Area (inch^2) = 1296t–0.00555

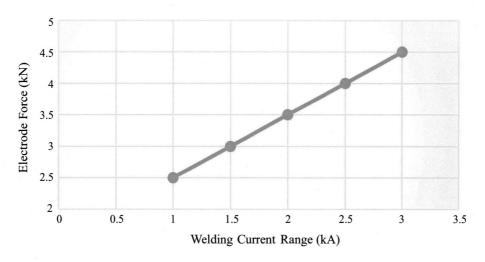

Fig. 4.13 Higher electrode forces expand the current range for advanced high-strength steels

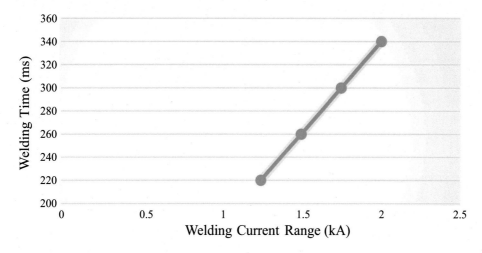

Fig. 4.14 Longer welding times expand the current range for advanced high-strength steels

- n = Nominal Nugget Diameter (inch) = $0.86\sqrt{1.65t - 0.007} = 0.86d$
- m = Minimum Nugget Diameter (inch) = $0.69\sqrt{1.65t - 0.007} = 0.69d$
- P = Welding Pressure (psi) = $60y_s + 10,200$
- F − Weld Force (pound) = AP
  - Weld Time for Bare Steel (cycles) = $0.017F + 2.5$
  - Weld Time for Galvanized Steel (cycles) = $0.017F + 7.5$.

Table 4.2 Adapted from AWS C1.1:2019 recommended practices for resistance welding. Spot welding parameters for bare and zinc-coated steel 350–700 MPa ultimate tensile strength*

Metal thickness (mm)	Electrode face diameter (mm)	Net electrode force (kN)	Coated weld time (mS)	Bare weld time (mS)	Approximate welding current coated (amps)	Approximate welding current bare (amps)
0.5	4.8	2.0	170	120	8,000	6,000
1.0	6.4	3.1	220	170	12,000	10,000
2.0	7.9	6.2	420	300	17,000	15,000

* *Note* This table of a random sampling of typical parameters serves as a rough guide and starting point only. Also, refer to the C1.1 standard for much more complete and comprehensive details

Table 4.3 Adapted from AWS C1.1:2019 recommended practices for resistance welding. Spot welding parameters for bare and zinc-coated steel > 700 MPa ultimate tensile strength*

Metal thickness (mm)	Electrode face diameter (mm)	Net electrode force (kN)	Coated weld time (mS)	Bare weld time (mS)	Approximate welding current coated (amps)	Approximate welding current bare (amps)
0.5	4.8	2.0	200	130	7,000	5,000
1.0	6.4	3.1	270	200	11,000	8,000
2.0	7.9	6.2	500	350	16,000	14,000

* *Note* This table of a random sampling of typical parameters serves as a rough guide and starting point only. Also, refer to the C1.1 standard for much more complete and comprehensive details

Either of these approaches can be used as a starting point when developing parameters for a given application.

4.5.2   Carbon Equivalence and Spot Weld Failure Modes

Because of their high alloy content, advanced high-strength steels tend to be much more hardenable (form martensite easily) relative to conventional steels. As mentioned in Sect. 4.3, a formula referred to as carbon equivalence (CE) is commonly used with conventional steels to assess the steel's hardenability. High CEs predict a high probability for martensite formation, and also indicate that the martensite that does form will be very hard. There are many CE formulas being considered for advanced high-strength steels, but their use and accuracy are not nearly as well established as they are for conventional steels.

But regardless of the formula used, carbon equivalencies for most advanced high-strength steels will be much higher than those of conventional steels, so the formation of

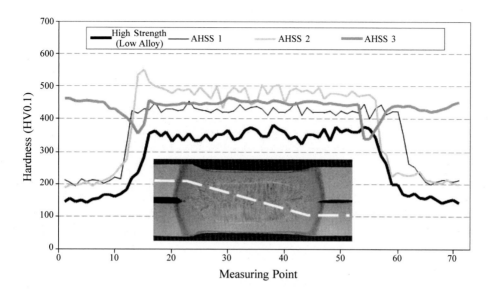

Fig. 4.15 Microhardness traverses of spot welds reveal much higher hardnesses for advanced high-strength steels compared to a conventional HSLA steel [*Source* WorldAutoSteel]

hard martensite when spot welding these steels is likely. For example, the WorldAutoSteel organization used a CE formula known as the Yurioka Equation to determine that CEs of some TRIP and DP steels tensile strength (in the range of 1,000 MPa) approached CEs of 0.6-an extremely high CE. Figure 4.15 compares microhardness traverses of three advanced high-strength steels to a conventional HSLA steel. Note that hardnesses of the advanced high-strength steels can be higher than the HSLA hardness by nearly 50% in this example.

Such high hardenability and hardness can be expected to affect spot weld failure modes during peel testing and chisel testing. A well-established industry standard associated with peel testing of conventional steels is that an acceptable test is one in which the peel test "pulls a nugget" or a "full button," as shown in Fig. 4.16. However, with advanced high-strength steels, full button pulls are less likely due to the high CEs that are likely to produce hard weld nuggets. This fact is compounded by the higher yield strengths of the material that will tend to produce greater stress concentrations at the edge of the nugget during a peel or chisel tests. Therefore, the conventional modes of testing such as peel and chisel testing may be more likely to produce interfacial or partial interfacial failure modes such as those shown in Fig. 4.17. With advanced high-strength steels, these types of failure modes may occur even though the weld strengths may be acceptable for the intended application. Even full interfacial failures may exhibit high strength.

The formation of high carbon hard martensite when arc welding steels has long been associated with a catastrophic form of cracking known as hydrogen cracking. This type

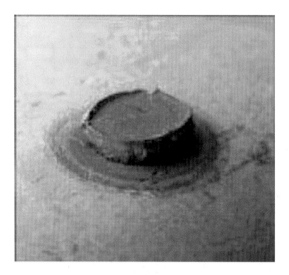

Fig. 4.16 The full button pull indicates acceptable weld quality when testing conventional steels [*Source* AWS]

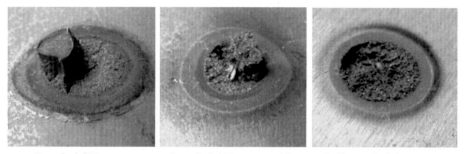

Interfacial Fracture   Interfacial Fracture with   Full Interfacial Fracture
with Button Pull   Partial Thickness Fracture

Fig. 4.17 Other button failure modes may be more likely with advanced high-strength steel spot welds due to higher weld hardness and higher base metal strength [*Source* AWS]

of cracking is especially dangerous since it occurs well after the weld has been made, and the weldment has cooled down. For this reason, it is often referred to as cold cracking or delayed cracking. So, whereas hydrogen cracking is a major concern when arc welding steels, it has not been a significant issue when spot welding conventional steels. However, one of the major factors promoting hydrogen cracking is a susceptible microstructure of hard martensite. And the higher the hardness of the martensite, the higher the susceptibility to hydrogen cracking. Since many advanced high-strength steels are known to form hard martensite during spot welding, it is entirely possible that hydrogen cracking may

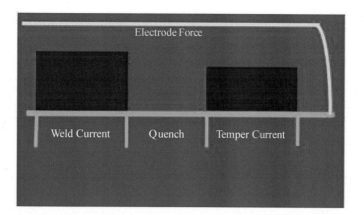

Fig. 4.18 The ability to conduct a tempering heat treatment during a single weld cycle is a rather unique capability of spot welding

become a concern with some of these steels, especially the higher strength versions. Currently, little is known about hydrogen cracking associated with spot welds, but with the rapid implementation of advanced high-strength steels it is likely to be an important topic of research in the immediate future.

But to improve nugget failure modes with advanced high-strength steels, the hard martensite must be softened. A simple and effective approach to accomplishing this is to add a temper cycle at the end of the weld (Fig. 4.18), something that can easily be easily added at the end of the spot welding cycle. Enough quench time must be included prior to tempering to allow the complete transformation to martensite, and the amount of tempering time and current will dictate how much softening occurs. Since the quench time adds cycle time to each weld, production requirements will mandate that this time be kept as short as possible. Depending on how hardenable the steel being welded, other approaches that slow cooling rates may be helpful. These approaches include current pulsing, longer weld times, and short hold times.

These techniques may be used to achieve better failure modes, but this will likely become more difficult to achieve with advanced high-strength steels that are among the higher-strength versions of these steels. So, while the standard practice with conventional steels has been to confirm weld quality through a full button pull, the "rules" may have to be altered when testing advanced high-strength steels. As indicated on Fig. 4.19, failure modes tend to go from full button pulls at low hardnesses associated with low-strength steels to interfacial failures at high hardnesses which can be expected when welding advanced high-strength steels.

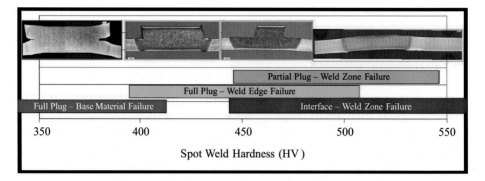

Fig. 4.19 Failure modes are related to weld metal hardness [*Source* WorldAutoSteel]

A better approach to developing spot weld schedules and assessing weld quality may be to combine a button fracture analysis with mechanical testing such as a tensile shear test (Fig. 4.20) and/or a cross tension test (Fig. 4.21). The cross tension is the more severe test of the two because it produces much greater stress concentrations at the edges of the nugget during loading. The higher the strength of the steel, the greater this effect will be. A ratio of cross tension strength to tensile shear strength is often referred to as spot weld "ductility" (or ductility ratio) and will reveal the extent to which the greater stress concentrations from the cross tension test affect the load at failure. A cross tension/shear tension ratio of close to one would be considered ideal, while numbers below 0.5 begin to indicate poor ductility. AWS D8.1, "Specification for Automotive Weld Quality Resistance Spot Welding of Steel" provides minimum test requirements for tensile shear and cross tension testing of advanced high-strength steels. Ultimately, the approach to testing of course will depend highly on the anticipated loading conditions in service. Mechanical property considerations of spot welds are discussed further in Chap. 7, and mechanical testing is covered in more detail in Chap. 8.

Fig. 4.20   Tension shear test

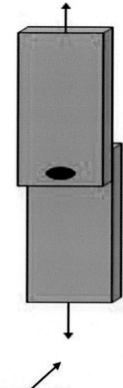

Fig. 4.21   Cross tension test

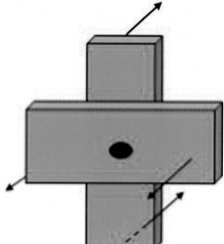

Resistance Spot Welding of Coated Advanced High-Strength Steels

5

5.1 Overview of Coated Steels

Most modern automobile components are very corrosion resistant, primarily due to the application of a zinc coating (prior to painting) which serves as a "sacrificial anode". This technique is known as cathodic protection which provides for continued protection of the steel even if some of the zinc is scratched off the surface. The most common approaches to applying zinc to the surface of steels are known has hot dipping and electro galvanizing. As indicated on Fig. 5.1, hot dipping is a lower cost process that is generally used on the lower-strength advanced high-strength steels. Electro-galvanizing (electro-plating) offers more precise control of the layer of zinc, and the ability to produce a thinner layer which improves weldability but sometimes at a greater cost. Most of the zinc-coated advanced high-strength steels are produced using the electro-galvanizing process. A more recently developed zinc coating includes small additions of magnesium to improve corrosion resistance in certain applications.

The galvanized steel may then receive a subsequent heat treatment producing what is known as a galvannealed coating (Fig. 5.2). The galvannealing process allows for diffusion to occur between the steel and the zinc, which creates layers of iron-zinc phases, and eliminates the pure zinc surface. Galvannealing produces a surface that is easier to spot weld and produces less electrode wear, but at the expense of long-term corrosion protection. The heat treatment also adds additional cost. Other metallic coatings such as aluminum and aluminum-silicon are sometimes used, but zinc coating is by far the most common and will be the focus of this section.

M. Kimchi and D. H. Phillips, *Resistance Spot Welding*, Synthesis Lectures
on Welding Engineering, https://doi.org/10.1007/978-3-031-25783-4_5

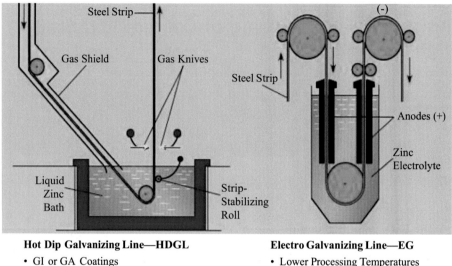

Fig. 5.1 The common approaches to galvanizing are hot dipping and electro-galvanizing [*Source* WorldAutoSteel]

Fig. 5.2 Galvannealing creates a zinc iron phase(s) that improves weldability [*Source* WorldAutoSteel]

5.2   Welding of Zinc-Coated Steels

Spot welding of zinc-coated steels necessitates much smaller process windows and longer weld times, and results in significant electrode wear as compared to welding uncoated steels. The accelerated electrode wear is due to a variety of reasons. As illustrated in Fig. 5.3, resistances are much lower with coated steels, meaning that much higher currents are needed which leads to hotter electrodes and faster wear. The reduction in resistance is due to several factors—less resistivity of the zinc, a softer surface which reduces contact resistance, and a larger current path created by a region of molten zinc around the nugget

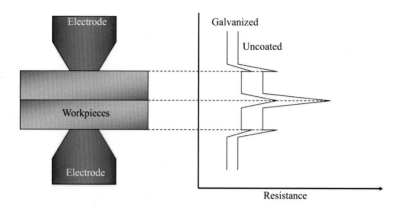

Fig. 5.3 Comparison of resistances during spot welding of galvanized versus uncoated steels

as it forms. Current levels as much as 50% higher than those used with uncoated steels may be required (Fig. 5.4).

The requirement for higher current is further accentuated by the fact that when welding galvanized steels, a single weld cycle usually requires at least two steps—the pure zinc coating must first be melted and pushed out of the way, and the weld nugget can then be formed. Figure 5.5 reveals both the significant increase in current needed, as well as the two stages of weld development—the initial melting and squeezing out of the zinc to allow steel-to-steel contact, followed by the formation of the weld nugget. Also notice that following steel-to-steel contact, the galvanized steel welds transition from initial weld

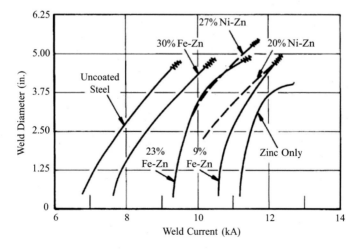

Fig. 5.4 Weld current requirements for uncoated steel, zinc-coated steel, and other various coatings (galvannealed and Ni-Zn) [*Source* The Ohio State University]

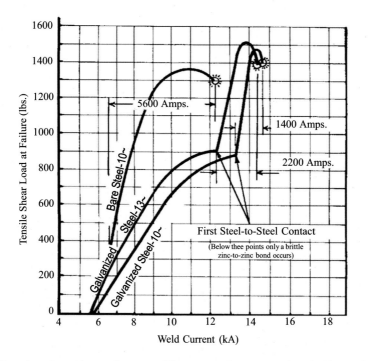

Fig. 5.5 Coated steels require much more welding current than uncoated steels (the numbers 10 and 13 indicate weld time in cycles) [*Source* WorldAutoSteel Steel]

nugget growth to expulsion over a range of 2,200 amps in one case (13 cycle weld time), and a range of 1,400 amps in the other case (10 cycle) weld. In comparison, the base steel 10 cycle weld has a current range of 5,600 amps-4 times the range of the 10 cycle galvanized steel weld. This indicates that one can anticipate much narrower process windows when welding galvanized steels (Fig. 5.6).

The narrow process window, combined with the requirement to make the weld in two steps, raises the possible concern of not reaching the second step. In this case, there is no weld even though it may appear that there is—the sheets are only held together by what is effectively a zinc brazed joint. This type of joint may actually perform well in tensile loading conditions, but the impact properties will be very poor. The additional step associated with welding galvanized steel also necessitates longer weld times. As indicated in Fig. 5.7, weld time increases as much as 50% or more may be needed when welding these steels.

The increased electrode heating due to the higher currents (and longer times) wears the electrodes through two mechanisms—"mushrooming" of the electrodes, and alloying of the copper electrodes with the zinc. These two mechanisms are shown in Fig. 5.8. Mushrooming of the electrodes occurs due to a combination of the reduction in yield strength as they get hot, and the clamping forces applied during welding. The formation of the

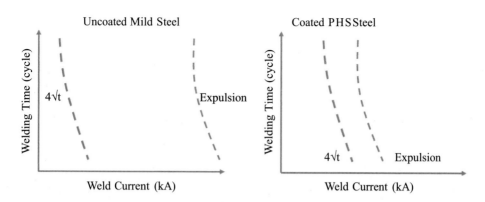

Fig. 5.6 Process windows are greatly reduced when welding coated steels
[*Source* WorldAutoSteel]

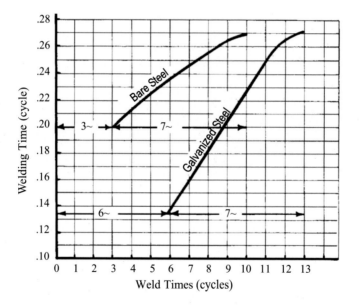

Fig. 5.7 Much longer weld times are needed when welding coated steels
[*Source* WorldAutoSteel]

zinc-copper phases occurs due to elevated levels of diffusion between the heated electrodes and the zinc coating. These phases are brittle and break away under the electrode pressure. Both mechanisms of electrode wear combine to cause the electrode diameter to rapidly increase during welding, which results in lower current density and smaller welds (Fig. 5.9).

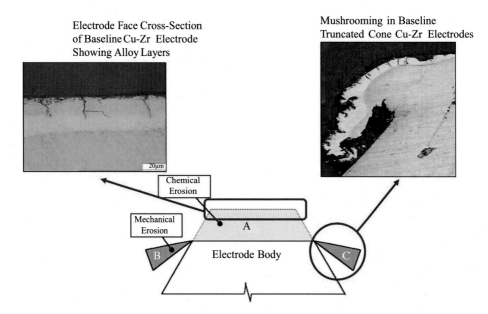

Fig. 5.8 The two mechanisms of electrode wear associated with spot welding of galvanized steels

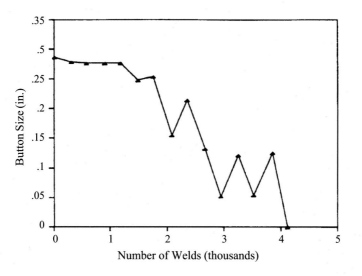

Fig. 5.9 Electrode wear when welding galvanized steels results in rapid drop in weld size
[*Source* EWI]

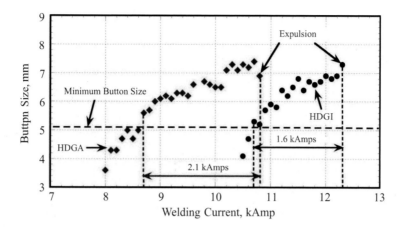

Fig. 5.10 Comparison of weld current requirements for a galvannealed (HDGA) DP steel and a galvanized (HDGI) DP steel reveal the benefits of galvannealing [*Source* WorldAutoSteel]

Thicker layers of zinc, which are sometimes required for even better corrosion protection, will accentuate the electrode wear problem, but as mentioned previously, galvannealed steels wear electrodes more slowly because they do not have a pure zinc surface. Figure 5.10 compares current levels (and ranges) for both a galvanized (HDGI) and a galvannealed (HDGA) DP steel. Significantly lower currents are needed for the galvannealed steel, which will obviously reduce the electrode wear. There is also an increase in the current range that will produce a wider process window, resulting in a more robust process. Other approaches to minimizing electrode wear when welding coated steels are discussed later in Chap. 9.

5.3   Welding of Coated PHS Steels

Spot welding of zinc-coated PHS steels is even more challenging since these steels are coated prior to the Hot Press Forming process which accelerates diffusion rates between the steel and the zinc coating. This results in alloying between the zinc and the steel and the subsequent formation of intermetallic layers, in particular a zinc-rich intermetallic known as Zn-Fe (Fig. 5.11). The figure shows a composition map for iron at the surface of a coated PHS steel, where the reddest shade indicates the highest concentration of iron. The α-Fe (Zn) is the region adjacent to the steel sheet that represents steel with dissolved Zn, while the adjacent orange layer is the Zn-Fe intermetallic, and an oxide (ZnO) is shown at the surface.

The development of a current range plot for a zinc-coated PHS steel using a typical single pulse weld cycle may produce expulsion before the minimum acceptable button diameter is even reached, meaning standard approaches to spot welding these steels may be nearly impossible. This is due to both the high resistivity and high melting point of

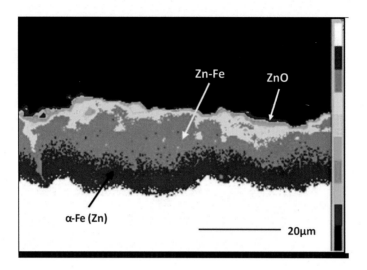

Fig. 5.11 The Zn-Fe intermetallic that forms during hot press forming zinc-coated PHS steels

the intermetallic. The WorldAutoSteel organization has reported significant improvements in welding zinc-coated PHS steels by using more complex weld schedules. In particular, a three-pulse schedule (Fig. 5.12) may significantly widen the weld process window. In this approach, the initial short high-current pulse breaks up the undesirable intermetallic layers to create good contact between the sheets. The second, longer, lower-current pulse initiates weld nugget formation while avoiding expulsion. The third, lower-current pulse grows the nugget to the required size. Figure 5.13 compares current range plots between a single-pulse weld and a multiple-pulse weld, clearly indicating the improvement in the process window with the multiple-pulse weld. However, it's important to note that such an approach adds time to the weld cycle and may not be desirable for high-speed production.

Because of the challenges associated with the zinc-coated PHS steels, many of the applications today utilize PHS steels with an Al-Si coating instead of zinc. Following the hot press forming operation, PHS steels coated with Al-Si coatings have been shown to be easier to spot weld because the intermetallics that form are more easily removed during welding. Figure 5.14 is an iron map comparing the faying surfaces (surfaces between the sheets prior to nugget formation) of an Al-Si coated PHS steel to a zinc-coated PHS steel. These surfaces were produced after three cycles of weld time, and clearly indicate that intimate contact between the steel sheets is achieved much more easily when spot welding Al-Si coated versus zinc-coated PHS steel. In fact, the weldability improvement is so significant that the current window when welding Al-Si coated PHS steels may be even wider than when welding PHS steels with no coating at all (Figure 5.15).

Figure 5.16 further illustrates the benefit of the Al-Si coating. As mentioned previously, the high resistivity and high melting point of the intermetallic that forms with Zn-coated PHS steels tends to cause expulsion prior before the nugget even begins to form. The

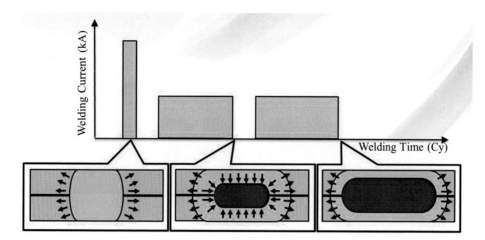

Fig. 5.12 Three-pulse weld schedule proposed for welding zinc-coated PHS steels

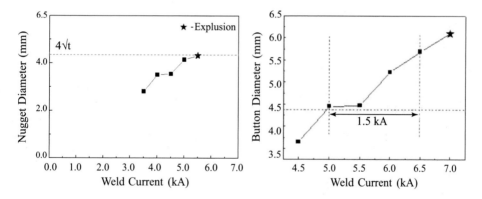

Fig. 5.13 Current range plot of a zinc-coated PHS steel-single pulse weld on the left, multiple pulse weld on the right

figure compares cross-sections of Al-Si-coated and Zn-coated PHS steels over 3 cycles of weld time. At 2 cycles, expulsion occurs with the Zn-coated steel but does not with the Al-Si steel. At 3 cycles, the previous expulsion leads to a poorly formed nugget Zn-coated steel nugget whereas the Al-Si coated nugget is properly formed.

Liquid metal embrittlement (LME), an embrittling phenomenon that occurs in the heat-affected zone of welds, is associated with the melting of a coating such as zinc, and the subsequent penetration of the molten coating into grain boundaries. This occurs at a very specific temperature range that is most conducive to the liquid material wetting and penetrating the boundaries. Once penetrated, the grain boundaries are susceptible to forming small cracks in the presence of tensile stresses surrounding the spot weld

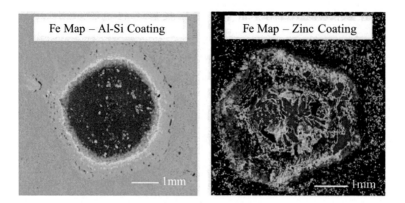

Fig. 5.14 Fe maps of faying surfaces after three cycles indicate much improved steel-to-steel contact when welding Al-Si coated PHS steels versus zinc-coated [*Source* Dong-Eui University]

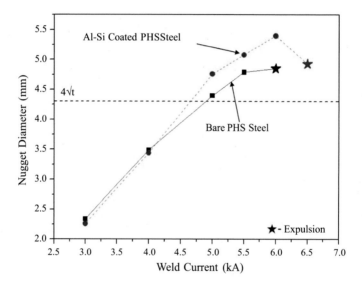

Fig. 5.15 Current range plots of an Al-Si coated PHS steel versus a bare PHS steel [*Source* Dong-Eui University]

(Fig. 5.17). Copper from the electrodes may also contribute to this form of cracking. Although this type of cracking has often been observed with spot welding, the possible negative effect of this cracking on mechanical properties, such as tensile and fatigue strength, has not been well established. LME may be more likely to occur with higher strength steels (i.e., above 800 MPa), and therefore, may become a more important issue as steel strengths continue to increase.

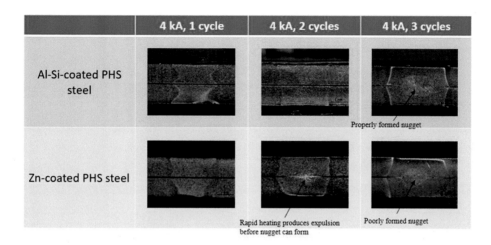

	4 kA, 1 cycle	4 kA, 2 cycles	4 kA, 3 cycles
Al-Si-coated PHS steel			Properly formed nugget
Zn-coated PHS steel		Rapid heating produces expulsion before nugget can form	Poorly formed nugget

Fig. 5.16 Initial expulsion with the Zn-coated PHS steel leads to a poorly formed weld nugget as compared to the Al-Si-coated steel

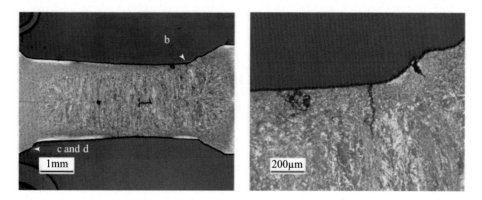

Fig. 5.17 Liquid metal embrittlement of a spot weld [*Source* GM and AET Integration]

There are many factors associated with production that can promote LME cracking. As shown in Fig. 5.18, typical factors that play a role are excessive weld time, worn electrodes, insufficient electrode cooling, excessive weld force, electrode alignment, and poor fit-up. LME cracking can be eliminated if all these factors are optimized (such as in a lab environment), but controlling these factors in production is difficult. Possible approaches to minimizing LME include the use of current pulsing to reduce heat input, more frequent electrode dressing, and longer hold times

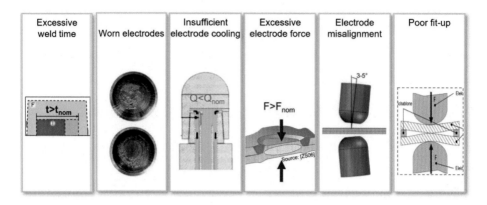

Fig. 5.18 Spot welding production factors that may promote LME cracking
[*Source* WorldAutoSteel]

Resistance Spot Welding of Aluminum

6

6.1 Overview of Aluminum

Because of its low density, high strength-to-weight ratio, and good corrosion resistance, aluminum offers many attractive features to automotive manufactures. In particular, the drive toward lighter-weight and more fuel-efficient automobiles is promoting greater use of aluminum. In fact, the use of aluminum in automobiles is expected to double by 2025. However, aluminum exhibits numerous properties and characteristics that present many unique spot welding challenges as compared to steel. The challenges associated with spot welding aluminum include a rapidly forming and tenacious oxide layer of variable thickness and composition, high electrical and thermal conductivities, small increases in resistivity with temperature, a narrow plastic range, low melting temperatures, and a high coefficient of thermal expansion. These issues will be discussed and recommended approaches to welding aluminum provided in this chapter.

6.1.1 Dynamic Resistance Curves

As indicated on Fig. 6.1, the dynamic resistance curve for aluminum is entirely different from the curve for steel. Two facts contribute to this vast difference-the oxide on the surface of the aluminum, and the small change in resistivity as a function of temperature. Upon initial flow of current, resistances are extremely high due to the oxide layer that has much higher resistivity than the aluminum. This increases the likelihood of initial expulsion and will also result in significant electrode heating. The oxide layer quickly breaks down, allowing current to pass more easily as resistance drops rapidly. However, as compared to the dynamic resistance curve for steel, there is no significant increase in resistance later in the cycle. Unlike steel, aluminum's resistivity increases only slightly

© The Author(s), under exclusive license to Springer Nature Switzerland AG 2023
M. Kimchi and D. H. Phillips, *Resistance Spot Welding*, Synthesis Lectures
on Welding Engineering, https://doi.org/10.1007/978-3-031-25783-4_6

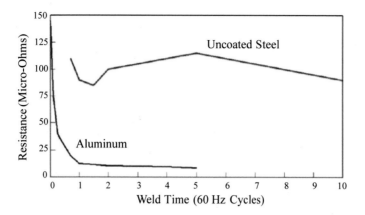

Fig. 6.1 Dynamic resistance curves for aluminum are much different than for steel [*Source* European Aluminum Association]

with temperature, as shown in Fig. 6.2. The implication of this difference is that there is limited opportunity to grow the nugget by taking advantage of the rapid increase in resistivity as is the case with steel. This is one of the fundamental reasons why aluminum spot weld times need to be much shorter than those of steel, as will be discussed later in this chapter.

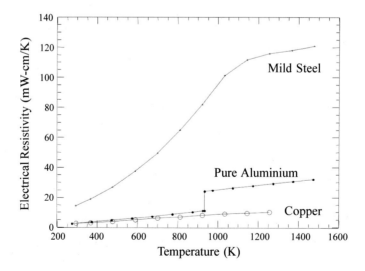

Fig. 6.2 The resistivity of aluminum increases only slightly with increasing temperatures [*Source* Resistance Welding Fundamentals and Applications]

6.1.2 Plastic Range

The plastic range of a metal can be loosely defined as the range of temperatures below its melting temperature in which the metal exhibits significant softening. The significance to spot welding is that wider plastic ranges will create a wider softened region around the weld for a longer time. This region, in conjunction with the electrode pressure, effectively "seals" the rapidly expanding (metals exhibit large volumetric expansion when they melt) molten weld nugget and prevents it from being ejected from the weld zone (expulsion). As indicated in Fig. 6.3, the typical plastic range of aluminum is significantly less than that of steel. The figure also includes a random heating line to illustrate the fact that a narrow plastic range not only reduces the width of the "seal" around the nugget, but also suggests that the window of welding time to produce a good weld is restricted. In summary, the narrow plastic range of aluminum, combined with its low melting temperature, means that the process window to create a good weld and avoid expulsion is very small.

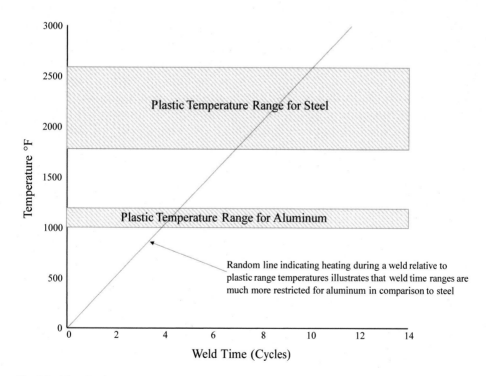

Fig. 6.3 The plastic temperature range for aluminum is much narrower than that for steel

6.1.3 Thermal and Electrical Conductivity

The RSW process works best with metal alloys such as steels that have electrical and thermal conductivities that are much lower than the copper-based electrodes used to weld them. Low electrical conductivity (or high resistivity) provides for easy I^2R heating, and low thermal conductivity means heat will be extracted from the weld nugget region more slowly. Aluminum exhibits electrical and thermal conductivities that are close to copper—two additional reasons contributing to spot welding challenges with this metal. These properties dictate the need for much higher currents, and much shorter times. Rules of thumb regarding welding currents and times are 2–4 times the current and 1/3 the time used when welding steel.

6.1.4 Coefficient of Thermal Expansion

The thermal expansion coefficient of aluminum is roughly three times higher than steel. This results in greater volumetric expansion of the metal upon heating, and subsequent greater contraction upon cooling. The consequence is a greater chance not only for expulsion, but weld discontinuities such as porosity and solidification cracking. This may mandate the need for low inertia, fast "follow-up" weld heads that can maintain consistent force during the rapid movement of the expanding and contracting weld region. Further improvements may be achieved by programming a current downslope at the end of the weld, and/or including a post weld forging step (increasing the force at the end of the weld).

6.1.5 Oxide Layer

As discussed previously, aluminum forms an oxide layer that forms rapidly and is tenacious. The highly resistive oxide layer can be a benefit in that it significantly increases contact resistance between the sheets being welded, but maintaining a consistent oxide layer thickness is difficult. On the other hand, if the oxide layer is significantly reduced by mechanical (such as grinding) or chemical (such as acid cleaning followed by a conversion treatment) methods immediately prior to welding, the need for extremely high currents will be mandated which will promote electrode sticking and accelerated wear. In some cases, aluminum suppliers may include a special chemical surface treatment to create a surface with higher resistivity than aluminum, but one that is more stable and consistent than the conventional oxide. The most practical approach may be for the suppliers to exercise special mill practice control to produce a uniform oxide layer, and for automotive manufactures to be diligent regarding shipping and storage practices.

6.2   Spot Welding of Aluminum

Because of the high electrical and thermal conductivity of aluminum, RWMA recommends Group A Class 1 electrodes which are more conductive than the Class 2 electrodes so common with spot welding of steel. A variety of electrode geometries may be used, but the Type "F" radius electrodes are generally the most common. As mentioned previously, much higher weld currents and shorter times are needed for aluminum as compared to steel—2–4 times the current and 1/3 the weld time. The typical aluminum lobe curve shape, shown in Fig. 6.4, illustrates the benefit of keeping weld times short (current range is widest here). But with relatively tight process windows and the propensity for expulsion, it may be helpful to include current upslope and downslope in the weld cycle if the machine has this capability. Table 6.1 shows the RWMA recommended practices when welding various thicknesses of aluminum using machines that have slope capability. Note that this table is based on 2XXX and 7XXX alloys which are more common in aerospace. But as the footnote indicates, this table can still be used as a guide for automotive alloys such as the 5XXX and 6XXX series, but somewhat lower values of time, current, and force should be used. Recommended practices for spot welding aluminum include an additional forge force (Fig. 6.5) which is important for reducing the likelihood of solidification cracking.

Recent studies in a report issued by the Aluminum Automotive Manual developed by the European Aluminum Association indicate that smaller electrode diameters with some electrode designs may be beneficial. Reduced electrode contact area creates more uniform contact distribution between the sheets and between the sheets and the electrodes, reduces intermittent undersize welds, and may increase electrode life. But regardless of the electrode type and design chosen, electrode wear when welding aluminum can be extremely rapid. This is due to the combination of the high heat generation at the electrode faces with the propensity for the aluminum to stick to the electrodes and begin to form copper/aluminum phases on the electrode face. These phases are brittle and exhibit high

Fig. 6.4 Typical aluminum lobe curve shape reveals the benefit of short weld times [*Source* European Aluminum Association]

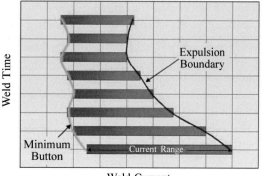
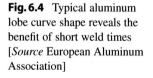

Table 6.1 Adapted from AWS C1.1:2019 recommended practices for resistance welding. Spot welding parameters for aluminum alloys on standard single-phase A-C type equipment*

Sheet thickness (mm)	Electrode face diameter (mm)	Top electrode radius (mm)	Net electrode force (kN)	Approximate welding current (amps)	Approximate welding time (ms)
0.5	16	1	1.5	18,000	80
1.0	16	3	2.7	31,000	130
2.0	22	4	3.8	42,000	170

*Note This table of a random sampling of typical parameters serves as a rough guide and starting point only. Also, refer to the C1.1 standard for much more complete and comprehensive details

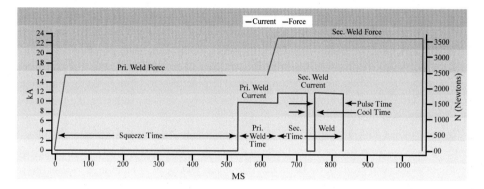

Fig. 6.5 Spot weld cycle for aluminum includes forge force

resistance, so once they begin to form electrode wear accelerates and weld quality rapidly deteriorates. Without proper electrode maintenance, as few as 300 welds can be expected from a set of electrodes.

The most efficient and practical approach to long electrode life and weld quality consistency when welding aluminum is to remove the aluminum that tends to stick and accumulate on the electrode surface. Standard practice has been to incorporate a regular tip dressing schedule into the spot welding production process. A variety of tip dressing equipment is available, ranging from hand-held devices for use with manual welding, to stand-mounted machines to be used with robots. These machines remove thin slices of the electrode face and generally maintain the electrode shape. Tip dressing is discussed further in Chap. 9.

However, the most important aspect to tip dressing electrodes when welding aluminum is not so much the need to maintain electrode shape and diameter, but to remove any aluminum from the electrode surface before it accumulates and deterioration becomes rapid. As reported in the Aluminum Automotive Manual, frequent and less aggressive buffing or grinding to remove aluminum from the electrodes early in the process results in much

Fig. 6.6 A simple and frequent buffing operation can be a very effective approach to long electrode life and consistent weld quality when welding aluminum [*Source* European Aluminum Association]

longer electrode life than standard tip dressing machines, which remove more material. The effect of periodic buffing after a relatively short number of spot welds has resulted in even more improved weld consistency and longer electrode wear. This approach is especially effective with aluminum because electrodes don't tend to mushroom like they do with steel, so a simple buffing operation to remove the aluminum just as it begins to form (Fig. 6.6) is all that is needed until electrode wear becomes significant.

Aluminum can be particularly susceptible to forming porosity in the spot weld nugget. This is due to the rapid increase in hydrogen solubility in liquid aluminum with increasing temperatures of the liquid. Combined with a major source of available hydrogen from moisture trapped in the oxide layer, liquid aluminum can easily dissolve a large amount of hydrogen as the nugget is forming. When the molten nugget begins to cool, the hydrogen comes out of solution in the form of a gas bubbles, which become trapped upon solidification, forming porosity. Small amounts of porosity (Fig. 6.7) will not adversely affect mechanical properties, and therefore, is typically deemed acceptable by the applicable code or specification. However, mechanical properties will certainly be degraded when porosity becomes severe (Fig. 6.8). The best approach to minimizing porosity is to control the aluminum oxide layer and maintain consistent welding conditions, which includes proper electrode dressing. Significant porosity is also more likely when heat input increases, such as when the electrodes begin to get hot and/or weld current or time is excessive. This places an even greater importance on keeping the electrodes cool and weld times short.

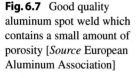**Fig. 6.7** Good quality
aluminum spot weld which
contains a small amount of
porosity [*Source* European
Aluminum Association]

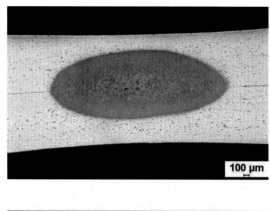

Fig. 6.8 Excessive amounts of
aluminum spot weld porosity
will result in mechanical
property degradation [*Source*
European Aluminum
Association]

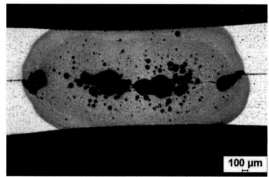

6.3   Novel Approaches to Spot Welding Aluminum

The problems associated with electrode wear when welding aluminum have driven
many organizations to pursue better ways to weld this light-weight metal. Two of those
approaches are summarized here.

6.3.1   GM's Multi-ringed Domed Electrode

GM has developed a spot welding process that utilizes a special domed electrode (Fig. 6.9)
that incorporates multiple concentric rings machined onto the electrode face. The purpose
of the rings is to break up the surface oxide, thereby reducing the contact resistance
between the electrodes and the sheets. This minimizes electrode heating and electrode
wear and allows for reduced amounts of current to be used due to an improved current
path through the sheets. It does require frequent dressing of the electrodes to maintain the
rings. GM has also reported wider process windows with this approach. Other approaches
studied to help break up the oxide include creating a rough surface on the electrodes.

Fig. 6.9 Three sheet aluminum spot welding utilizing GM's multi-ringed domed electrode [*Source* GM]

6.3.2 Fronius DeltaSpot Process

Fronius offers a unique spot welding approach called DeltaSpot which utilizes an intermediate layer (process tape) between the electrode and the aluminum sheet (Fig. 6.10). The tape is continuously fed at a speed that is coordinated with each spot weld, producing a fresh section of tape with each new weld. Process tapes are available in a range of different alloys and coatings, with different electrical and thermal conductivities. The process tape prevents direct contact between the electrode and the aluminum sheets which greatly reduces electrode wear. In addition, since each weld is made with a fresh tape surface, more consistent welds can be made with a wider process window. Most important, however, is the possibility to control the heat balance more precisely in the work piece as compared to standard spot welding. This is due to the additional resistances associated

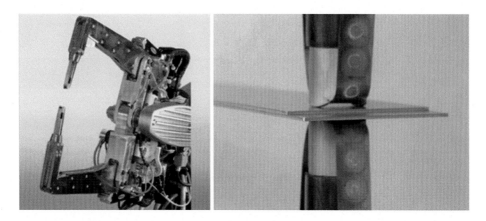

Fig. 6.10 The Fronius DeltaSpot Process utilizes a continuously fed process tape that produces additional heating and separates the electrodes from the aluminum sheets [*Source* European Aluminum Association]

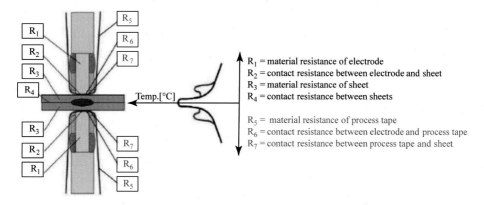

Fig. 6.11 The Fronius process tape provides additional heat balance control due to the three additional resistances (R5, R6, and R7) [*Source* European Aluminum Association]

with the tape (Fig. 6.11), which allows for the possibility of customization of the heat balance through adjustments to the tape's composition and/or coating.

6.3.3   Refill Friction Stir Spot Weld (RFSSW) Process

Whereas the RFSSW (also referred to as SpotMeld) process (Fig. 6.12) is a solid-state friction welding process and not a resistance welding process, it's configuration is similar to spot welding, and therefore adaptable to automotive assembly lines. The various steps

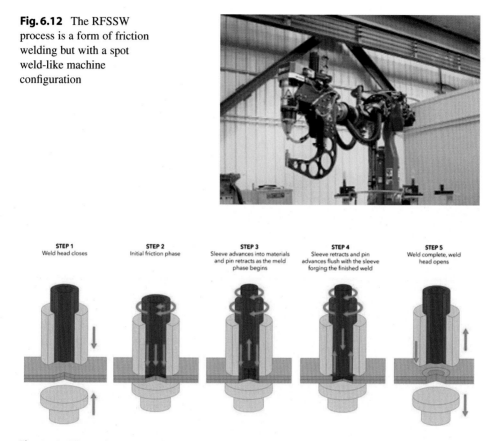

Fig. 6.12 The RFSSW process is a form of friction welding but with a spot weld-like machine configuration

Fig. 6.13 The various steps of the RFSSW process

associated with the process are shown in Fig. 6.13. The initial friction phase begins to soften the top sheet due to frictional heating. The so-called "meld phase" continues to develop more softened aluminum from the bottom sheet which is drawn up into the cavity created by the tool. The final forging step creates the shape of the weld which is flush with the surface. Also notice that the weld does not penetrate all the way through the second sheet offering the benefit of a completely smooth surface on that side.

Joining Aluminum to Steel—Spot Welding and Other Approaches

7

7.1 Background

The strong interest in reducing the weight of automobiles through the use of lighter weight materials such as aluminum inevitably results in the need to weld aluminum to steel. Unfortunately, these metal alloys are metallurgically incompatible, and therefore extremely difficult to weld with processes that involve fusion (melting). The incompatibilities that make welding them difficult include significant differences in melting temperature, forging temperature, and coefficient of thermal expansion. A summary of the differences in properties is shown in Table 7.1. When mixed, they will also form brittle intermetallics. So, to date, most of the successes in joining these two metals has been with solid-state welding processes such as explosion welding. But this section mainly focuses on spot welding and approaches representing a similar configuration to spot welding, and therefore offer the most promise for an automotive assembly line.

7.2 Spot Welding

Attempting to spot weld these two metals will most certainly result in the formation of brittle intermetallics. In fact, a glance at the Al–Fe phase diagram reveals the likelihood of multiple intermetallics forming when aluminum and steel are mixed. So, while it may be possible to produce a fusion weld such as a spot weld that exhibits some tensile strength, the fracture toughness will be extremely low and the likelihood of cracking high. Figure 7.1 shows a region of a spot weld between a zinc-coated steel and aluminum and a hardness plot across the interface. As the plot clearly indicates (as do the size of the indents on the picture), the Al–Fe intermetallic layer in the center exhibits extremely high hardness which would certainly coincide with poor toughness.

© The Author(s), under exclusive license to Springer Nature Switzerland AG 2023
M. Kimchi and D. H. Phillips, *Resistance Spot Welding*, Synthesis Lectures on Welding Engineering, https://doi.org/10.1007/978-3-031-25783-4_7

Table 7.1 Physical property comparison between Al and Fe

Properties	Al	Fe
Specific heat (J kg^{-1} K^{-1})	1080	795
Melting point (°C)	660	1536
Thermal conductivity (W m^{-1} K^{-1})	94	38
Thermal diffusivity (m^2 s^{-1})	3.65×10^{-5}	6.8×10^{-6}
Coefficient of thermal expansion (1/K)	24×10^{-6}	10×10^{-6}
Density (kg/m^3)	2385	7015

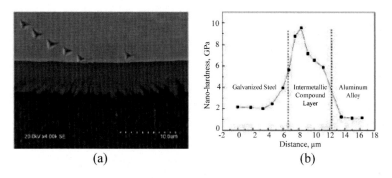

(a)   (b)

Fig. 7.1 Al–Fe intermetallic (center layer in micrograph) is extremely hard and brittle [*Source* Key Laboratory of Automobile Materials, School of Materials Science and Engineering, Jilin University]

   To avoid or minimize the formation of brittle intermetallics, the use of interlayers has been explored. Types of interlayers that have been studied include aluminum cladded steel (achieved by cold rolling) sheets, austenitic stainless steel, and various alloys of aluminum. Some success regarding tensile strength achieved with some of these approaches has been reported, but the implementation of interlayers in a high production spot welding environment does not appear to be practical.

7.3   Unique Welding Approaches with Potential for Use on Automotive Assembly Lines

7.3.1   Spot Welding with Rivet-Like Insert

One approach that has been developed is very similar to spot welding, but with a steel rivet inserted into the aluminum sheet. As indicated in Fig. 7.2, a hole is drilled into the aluminum sheet in which a rivet-like steel plug is placed, and then current is passed

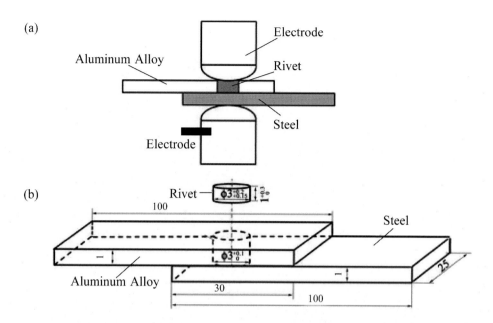

Fig. 7.2 An approach to spot welding aluminum to steel utilizing a rivet-like insert [*Source* School of Materials Science and Engineering, Henan University of Science and Technology, Luoyang, China]

through the entire assembly to create the weld. Although intermetallic formation was again identified, some tensile strength improvement as compared to direct spot welds between aluminum and steel was reported. However, the introduction of an additional drilling and rivet placement operation would likely add considerable time and expense to a typical spot welding line.

7.3.2 Flexweld

An adaption of the rivet-like insert approach is now being utilized by Volkswagen and is known as Flexweld (Fig. 7.3). Flexweld is a two-step process in which a steel insert (called a steel resistance element) is first punched into the aluminum sheet, and then the insert is welded directly to the steel sheet to produce the joint. An adhesive is also included to add additional strength. Because the hole in the aluminum is created by the placement of the insert in a single step, the entire process is easily automated, and Volkswagen is currently using it on some of their production lines. Whereas the process utilizes typical resistance spot welding equipment and electrodes, the weld that is produced is more similar to a projection weld.

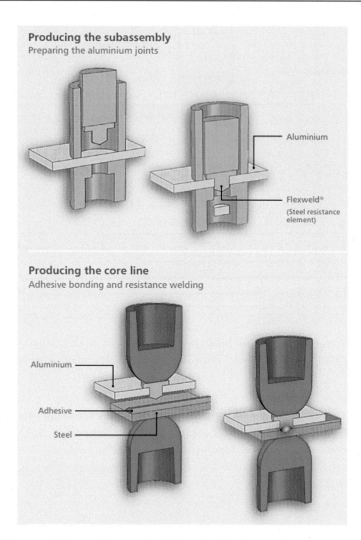

Fig. 7.3 Volkswagen's Flexweld process for joining aluminum to steel

7.4   Mechanical Methods for Joining Aluminum to Steel

Because of the metallurgical incompatibilities that make it so challenging to weld aluminum to steel, mechanical methods for joining them may be an option. There are many types of mechanical joining methods, but the two that are most used in the automotive industry are clinching and self-pierce riveting. The main advantages of mechanical joining processes are high tool life, simple equipment, and relative shorter process times.

Drawbacks include the requirement for access from both sides, bulges and indents, and relatively high amounts of force required to make the connections.

7.4.1   Clinching

Clinching is a process that includes offsetting, upsetting and flow pressing. As shown in Fig. 7.4, clinching is a punch and die process that causes the materials to flow to create a mechanical interlock. Since the process relies on material flow (or plastic flow), higher strength steels may be more difficult to clinch to aluminum because of the increased difficulty of creating plastic flow which will produce less of an interlock. Clinching should only be considered for non-critical applications.

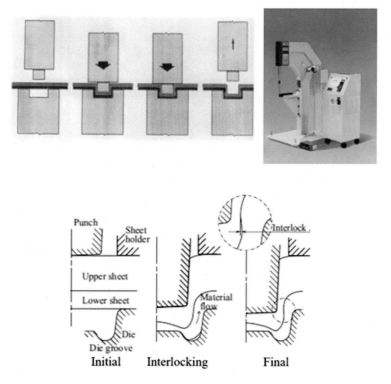

Fig. 7.4   The clinching process relies on material flow to create an interlock

7.4.2   Self-Pierce Riveting

Self-pierce riveting (SPR) is a mechanical joining process in which two or more sheets are joined by inserting a rivet from the top sheet(s) into the bottom sheet(s), and then locking it in place by pressing it into a die. It is a widely used mechanical joining process for joining steel to aluminum. As shown in Fig. 7.5, it consists of the four stages of clamping, piercing, flaring, and releasing. The main advantages are no predrilling or surface treatment is needed, and there are no sparks or fumes. Another important advantage of self pierce-riveting is improved fatigue performance (Fig. 7.6) over spot welds. This is due to the elimination of the sharp notch associated with the outer edge of a spot weld nugget. However, as shown in Fig. 7.7, the shear strength of spot welds is much higher.

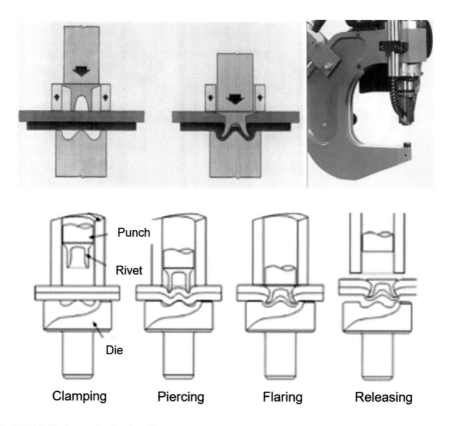

Fig. 7.5 Self-pierce riveting is a four-step process

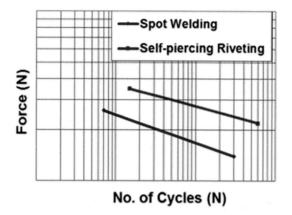

Fig. 7.6 Comparison of spot weld and self-piercing riveting fatigue strength

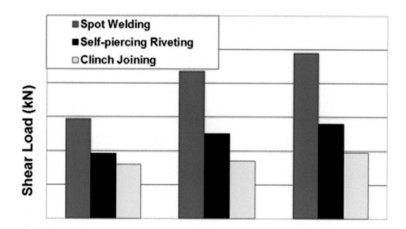

Fig. 7.7 Comparison of spot weld, self-piercing riveting, and clinch joining shear strength

Design and Weld Strength Considerations for Resistance Spot Welding

8

8.1   Spot Weld Mechanical Property Considerations

The strength of a spot weld is a function of its size. This is the reason nugget size plays such an important role in design criteria and the development of Lobe curves. However, it's important to point out that spot welds are not very effective at supporting loads. This is demonstrated in Fig. 8.1, which is a plot of tensile shear strength (load) as a function of spot weld nugget size for various thicknesses of a plain carbon conventional steel. Although the plot does establish the relationship between strength and nugget size, the increase in load carrying ability of the spot weld increases only slightly with increasing nugget diameters. For example, an increase in the nugget diameter shown for the 1.91-mm thick sheet from 0.2 to 0.4 inches equates to an increase in nugget area almost fourfold, but the increase in the load it can carry is only about 1.5 times.

The reason for the poor mechanical performance of spot welds is they do not evenly distribute loads since the stresses will always be concentrated along the outer diameter of the nugget. Some loading conditions will produce even more severe stress concentrations than a tensile shear condition and will result in lower failure loads. Figure 8.2 compares three common loading conditions/test types for a spot weld—tensile shear, cross tension, and peel/chisel. As the figure indicates, the severity of the loading condition is the least in tensile shear and the greatest in a peel or chisel condition. This figure also reveals that a spot weld loaded in shear will perform better in a mechanical test, such as tension than one that is in a peel-type loading condition. Designers must compensate for these deficiencies in spot weld mechanical performance by producing a sufficient number of welds, and when possible, utilizing designs that place welds under loading conditions that best distribute the load.

In addition to the loading condition, it is also important to consider that there may be some softening in the region of highest stress. Notice on Fig. 8.3 (a figure shown

© The Author(s), under exclusive license to Springer Nature Switzerland AG 2023
M. Kimchi and D. H. Phillips, *Resistance Spot Welding*, Synthesis Lectures
on Welding Engineering, https://doi.org/10.1007/978-3-031-25783-4_8

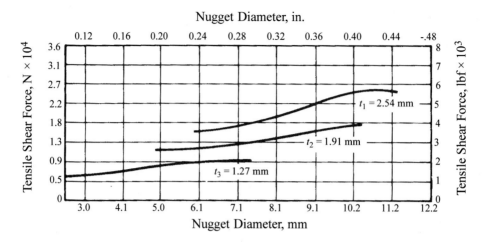

Fig. 8.1 Spot weld tensile shear strength as a function of nugget diameter [*Source* ASM Handbook Volume 6]

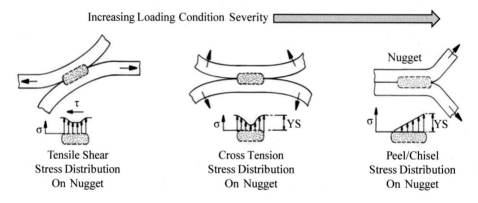

Fig. 8.2 The effect of loading condition on nugget stress distribution

previously, Chap. 4, Fig. 4.15), the advanced high-strength steel labeled AHSS 3 shows a softened region at the weld edge. In this case, the softened region is due to the heat from the weld tempering the base metal martensite. When welding a material that was strengthened by strain hardening (cold working), a similar softening can be anticipated due to the likelihood of recrystallization. These regions could promote desirable peel test results based on the visual of a pulled nugget, and at the same time, failures at very low loads due to the softened regions being so close to the regions of stress concentration during loading.

Depending on the steel alloy being welded, there may be other mechanisms associated with softening outside of the nugget. HF steels may form a softened region at the fusion

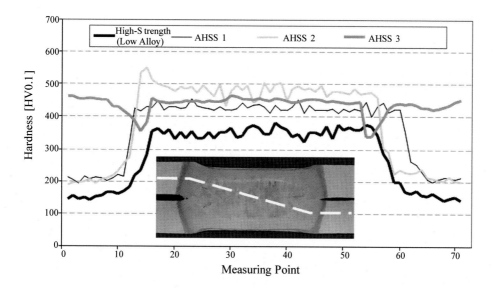

Fig. 8.3 Sample AHSS 3 exhibits significant base metal softening at the edge of the nugget [*Source* WorldAutoSteel]

line (Fig. 8.4) which has been associated with the formation of a carbon depleted region (Fig. 8.5). Researchers at Otto von Guericke University Magdeburg, Institute of Materials and Joining Technology proposed that the formation of this region occurred when a narrow region at the fusion line was heated into the delta phase and then rapidly cooled. During this period, carbon migrated out of the delta phase (which can dissolve much less carbon than austenite) and didn't have sufficient time to diffuse back in upon cooling. This softened region was eliminated by going to a much shorter weld time. However, probably the most important observation was that while the welds with the softened region promoted more of a full button mode failure (vs. an interfacial failure with the welds that did not have softening) in the cross tension test, it didn't significantly impact the overall tensile strength.

Researchers at Ohio State identified a softened region of an HF steel farther away from the fusion line (Fig. 8.6). This region is sometimes referred to as the "subcritical heat-affected zone" and was attributed to the over tempering of martensite which produced a soft region of ferrite with carbides.

8.2 Multiple Sheet Stack-Ups

Current automobile designs rely heavily on components that offer highly customized properties across the component. For example, a door frame may consist of some steels that are extremely high strength to resist an impact, and others that are much lower strength

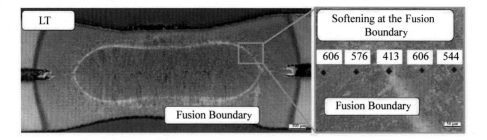

Fig. 8.4 Softened region along the fusion line of an HF steel [*Source* Otto von Guericke University Magdeburg, Institute of Materials and Joining Technology]

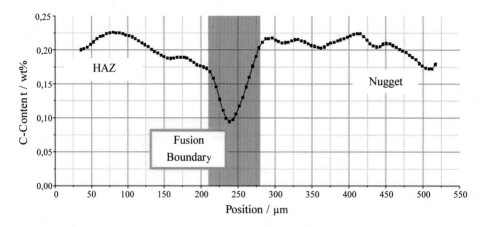

Fig. 8.5 Softened region along the fusion line of HF steel is attributed to a depletion of carbon in this region [*Source* Otto von Guericke University Magdeburg, Institute of Materials and Joining Technology]

to help create a crumple zone. For the same reason, thickness within a component may be varied as well. To produce such customized components, multiple sheet stack-ups are often needed. This usually involves producing a single spot weld to weld together 3 or 4 sheets which often have varying thicknesses. The configuration of a given stack-up will of course determine the welding approach to be used, but it would be well beyond the scope of this book to try to address all multiple sheet stack possibilities. Instead, a couple of examples will be discussed that should be helpful when determining an approach to any stack-up design.

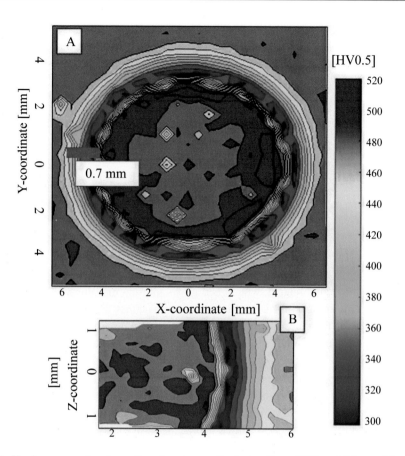

Fig. 8.6 Hardness map showing softened region on the fusion line of HF steel [*Source* The Ohio State University]

The first example is a three-sheet stack up in which the bottom and middle sheets are of similar thickness and the top sheet is much thinner. Attempting to spot weld this arrangement using a conventional weld cycle will result in difficulties creating a weld between the top sheet and middle sheet due to the electrode cooling effect at the much thinner top sheet. A two-step approach that was developed to overcome this problem is shown in Fig. 8.7. The first step utilizes a short pulse of high current combined with low electrode force. The very short times do not allow enough time for the electrode cooling to dominate at the top sheet so a weld can easily form between the top and middle sheet. The low force creates high contact resistances between the sheets. The combination of low force and short times at high current produce a weld that relies mainly on high contact resistance and less on bulk resistance of the sheets. Because the top sheet is thin, heating from contact resistance is rapid and a nugget forms first here. Once that nugget has formed, step two utilizes a more conventional lower force, lower current, and longer

Fig. 8.7 The ability to vary both current and force is an effective approach to this three-sheet stack-up [*Source WorldAutoSteel*]

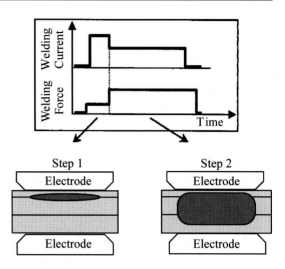

time to form the second nugget, which then grows into the first nugget to form a single weld nugget. This approach of course is limited to machines that are capable of varying force during the weld cycle, but does illustrate the potential benefits associated with this additional machine capability.

Another example is a three-sheet stack-up in which the two outer sheets are thinner than the middle sheet. In this case, as reported by the WorldAutoSteel organization, two multiple-pulse versions of varying the current magnitudes and times were shown to be effective approaches (Fig. 8.8). When welding a three-sheet stack-up in which all three sheets are of different thicknesses, similar approaches proved effective. A typical example with the weld produced is shown (Fig. 8.9). In summary, the common theme in all examples of three-sheet stack-ups is a two-step process which begins with a high current pulse followed by a low current pulse or pulses, however, in some cases a lower initial pulse may be helpful. These more complex pulsing schedules must be developed experimentally, and potentially with computational modeling support. In addition, the option to vary force during the weld cycle provides a further dimension of control that may prove even more valuable in more complex stack-ups than three sheets. One more consideration that may be beneficial in some cases is to vary the top and bottom electrode face diameters such that the smaller electrode face is on the thinner material side of the stack-up.

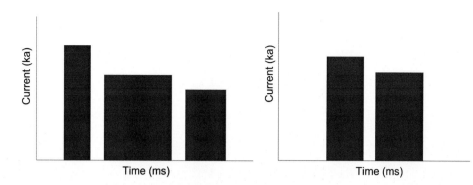

Fig. 8.8 These approaches were shown to be effective for a three-sheet stack-up made up of two outer thin sheets and a thicker middle sheet [*Source* WorldAutoSteel]

Fig. 8.9 A slight variation from the approach in Fig. 8.8 was effective when all three sheets were of different thickness [*Source* WorldAutoSteel]

8.3 One-Sided Welding

In some cases, it is not possible to gain access to both sides of the component being spot welded, mandating the need for one-sided welding, also known as indirect spot welding. Figure 8.10 shows a typical one-sided spot welding set-up. Two electrodes are placed in parallel above the part to be welded. Notice that in this case one of the electrodes is a larger flat electrode, referred to as a shunt electrode. The shunt electrode passes current without generating significant heat allowing for the conventionally shaped electrode next to it to produce a single weld. The copper base shown on the figure provides a good heat sink. By replacing the shunt electrode with a second conventionally shaped electrode two welds can be produced instead of one.

The WorldAutoSteel organization reported on a successful one-sided welding approach which involves two current steps combined with the ability to vary force during the weld cycle (Fig. 8.11). As illustrated in the figure, numerical simulations were also developed which reveal the heat generation and heat flow patterns occuring during this weld. The initial stage of low welding current and high electrode force preheats the sheets which

Fig. 8.10 One-sided spot welding set-up will produce a single weld under the conventionally shaped electrode (on the right) [*Source* EWI]

increases the contact area between the electrode and upper sheet, and thus, inhibits surface expulsion. This increased contact area also decreases the current density through the upper sheet, while at the same time, the contact area between upper and lower sheets is increased, forming a more stable conduction path. A lower second stage force avoids increasing contact area any further, and then a relatively long higher current pulse produces the weld.

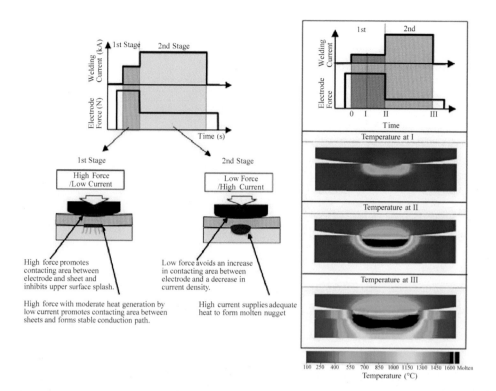

Fig. 8.11 Approach to one-side welding (only one of two electrodes is shown) utilizes low current and high forces in the first step, followed by high current and low forces in step 2 [*Source WorldAutoSteel*]

8.4   Weld Bonding

Weld bonding refers to a common automotive manufacturing process in which spot welds are made between two sheets after an adhesive has been applied. It is beyond the scope of this book to cover this process in detail, but a summary is provided here. As discussed earlier in this chapter, individual spot welds are not very effective at supporting mechanical loads because of the stress concentrations that are created at the edge of the nugget. The addition of an adhesive between the sheets provides much better load distribution, an especially important concept when fatigue loading conditions are expected. Typically, the adhesive bond is designed to carry the majority of the load, so usually the primary purpose of the spot welds is to simply hold the sheets together to allow the adhesive time to cure. When producing the spot weld, the electrode force is sufficient to squeeze the uncured adhesive out of the way prior to forming a nugget. Curing of the adhesive then occurs after the spot weld is made and may or may not require additional heating in the oven. One possible problem that can occur is if the adhesive cures prior to making the spot weld. This will not allow the electrode force to properly squeeze out the adhesive, and a poor weld with excessive expulsion will result.

Resistance Spot Welding Quality, Testing, Monitoring, and Control

9.1   Discontinuities and Defects

There are many possible spot weld defects and discontinuities that can occur for a wide variety of reasons, but the most common reasons are contaminated parts, improper welding parameters and/ or electrode geometry, poor part fit-up, and electrode wear. Whether or not the discontinuity (or weld imperfection) is considered a rejectable or unacceptable defect is dictated by the applicable welding code (Chap. 10) or company specification. Figure 9.1 provides a summary of several common spot welding defects/discontinuities and possible causes. For more extensive coverage of this subject, refer to AWS C1.1, "Recommended Practices for Spot Welding."

9.2   Destructive Testing of Spot Welds

A wide variety of destructive tests are available for testing spot welds. Common examples include peel, chisel, bend, tension-shear, cross-tension, u-specimen tension, impact, and fatigue. All these approaches and more are described in detail in AWS C1.1, "Recommended Practices for Resistance Welding". This section will briefly review some of the most common approaches, with emphasis on testing considerations for modern automotive materials.

9.2.1   Peel and Chisel Testing

The two most common, simple, and inexpensive ways to test a spot weld are the peel test and chisel test (Fig. 9.2). In both cases, the sheets are forcibly separated until failure occurs, either in the surrounding base metal or the nugget itself. The tests may be

M. Kimchi and D. H. Phillips, *Resistance Spot Welding*, Synthesis Lectures on Welding Engineering, https://doi.org/10.1007/978-3-031-25783-4_9

Condition   Likely Causes

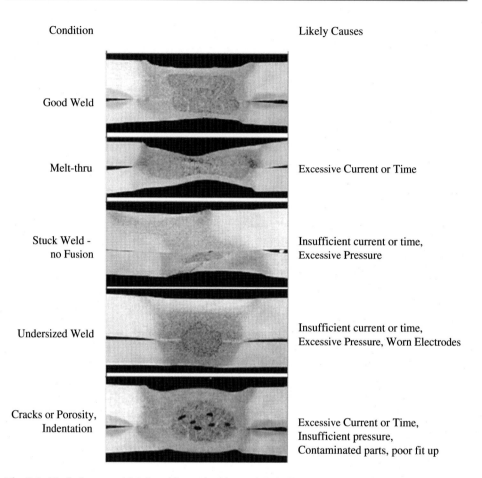

Good Weld

Melt-thru   Excessive Current or Time

Stuck Weld -   Insufficient current or time,
no Fusion   Excessive Pressure

Undersized Weld   Insufficient current or time,
                                                             Excessive Pressure, Worn Electrodes

Cracks or Porosity,   Excessive Current or Time,
Indentation   Insufficient pressure,
                                                             Contaminated parts, poor fit up

Fig. 9.1 Typical spot weld defects/discontinuities and their likely causes

used to determine whether the nugget is the proper size, or if it is of acceptable quality (as indicated by the failure mode). The peel test may be instrumented to provide the load at failure. Such tests have been used extensively over the years with conventional steels where it is a commonly accepted practice to base weld quality purely on whether a full button pull is achieved. However, as discussed previously, testing has shown with modern high-strength materials, such as advanced high-strength steels, that fracture modes (Fig. 9.3) other than a full-button pull may produce acceptable mechanical properties. Automotive designers and engineers may need to consider more thorough and/or innovative approaches to testing spot welds of advanced materials to develop spot weld procedures and assess weld quality. In particular, the ability to monitor loads at failure with any chosen test in many cases may prove to be useful.

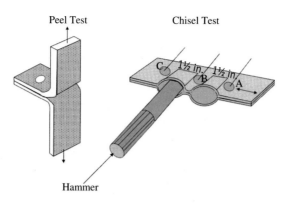

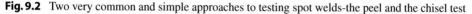

Fig. 9.2 Two very common and simple approaches to testing spot welds-the peel and the chisel test

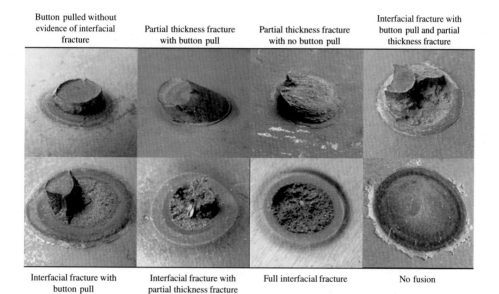

Fig. 9.3 A wide variety of spot weld fracture modes are possible, and modes other than a full-button pull (upper left) may be acceptable with higher strength materials [*Source* WorldAutoSteel]

One novel approach for consideration is a version of the chisel test known as wedge testing (Fig. 9.4), which was recently studied at The Ohio State University. The test utilized a tensile testing machine to provide the necessary force, and a camera and digital imaging software were used to study the strain during loading (Fig. 9.5) using an approach known as digital imaging correlation (DIC). A high-strength (1500 MPa) Boron steel with a range of weld sizes was studied. While this approach to testing may be too complex and

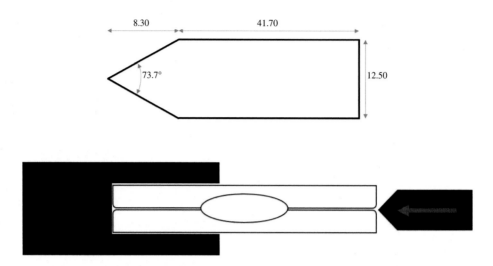

Fig. 9.4 The wedge test [*Source* The Ohio State University]

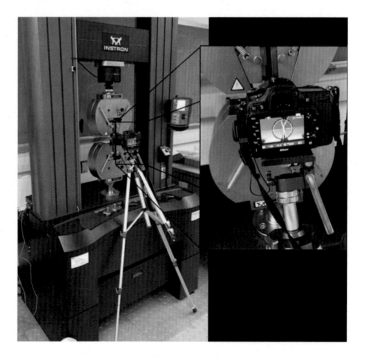

Fig. 9.5 The wedge test combined with DIC can provide a variety of information (such as strain patterns) that can be used in the development of computational models [*Source* The Ohio State University]

time consuming for a production environment, it is an approach that could be used if a thorough understanding of the strain patterns and associated failure modes is needed. One of the many conclusions from this study was that smaller weld sizes failed along the weld interface with minimal strain, while large sizes failed in the base metal with significant strain. Figure 9.6 shows actual strain patterns of three different weld sizes. While these results would not be considered surprising, much of this information could be useful in developing numerical modeling software for studying the mechanical behavior of spot welds.

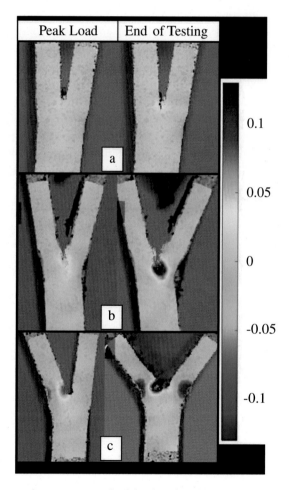

Fig. 9.6 Wedge test strain patterns reveal minimal strain with the small weld (top of figure), and maximum strain with the largest weld (bottom of figure) [*Source* The Ohio State University]

9.2.2 Mechanical Testing

Tension-shear and Cross-tension Testing

Tension-shear (Fig. 9.7) and cross-tension (Fig. 9.8) testing are two of the most common mechanical tests for spot welds. As reported by the WorldAutoSteel organization, these tests are becoming more important for testing higher strength materials due to the decreasing emphasis on relying solely (or primarily) on a fully pulled nugget to define an acceptable weld. Figure 9.9 shows tensile shear spot weld test results of a TRIP and DP steel, as well as two more conventional steels. These results indicate that weld tensile strength is proportional to base metal tensile strength, which is desirable. When testing thicker advanced high-strength steel spot welds (ranging from small button size to expulsion button) the fracture mode during shear-tension testing may change from interfacial to button pull out. But even the interfacial failure modes may demonstrate high load-bearing capacity. However, in thinner gage steels, an acceptable fracture mode is often a button.

In a recently published study reported by the WorldAutoSteel organization, Finite-Element Modeling (FEM) and fracture mechanics calculations were used to predict spot weld fracture mode and loads in tension-shear tests of advanced high-strength steels. The results were compared to those obtained for conventional steels. The results of the work confirmed the existence of a competition between two different types of fracture modes-the full button pull and the interfacial fracture. The force required to cause a complete weld button pull fracture was found to be proportional to the tensile strength and the thickness of the base metal, as well as the weld diameter. The force to cause an interfacial weld fracture was related to the fracture toughness of the weld, the sheet thickness, and the weld diameter. It was determined that there is a critical sheet thickness above which the expected fracture mode could transition from a full button pull to an interfacial fracture. This analysis revealed that as the strength of the steel increases, the fracture toughness of

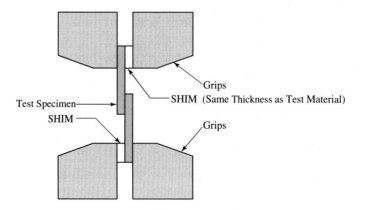

Fig. 9.7 Tension-shear test

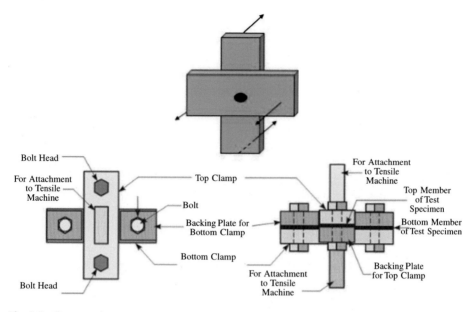

Fig. 9.8 Cross-tension test

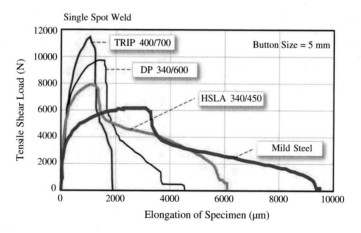

Fig. 9.9 Tensile shear strength of single spot welds [*Source* WorldAutoSteel]

the weld required to avoid an interfacial fracture must also increase. Therefore, despite higher load-carrying capacity of advanced high-strength steels, their spot welds may be prone to interfacial fractures. Tensile testing showed that the load-carrying capacity of the samples that failed via interfacial fracture was found to be more than 90% of the load associated with a full button pull. This indicates that the load-bearing capacity of the welds is only slightly affected by the fracture mode, meaning load-bearing capacity, not fracture mode, should be the primary method to evaluate spot welds in advanced high-strength steels.

Impact Testing

Impact testing is very different from tensile testing in that it evaluates the ability of the spot weld to resist fracture under impact type loading conditions, such as those that occur during a car crash. As described in AWS C1.1, there are many types of impact tests that can be used for spot welding. The test shown in Fig. 9.10 is known as the cross-tension drop-impact test. This test utilizes calibrated springs (or a force transducer) and a free-falling weight. If calibrated springs are used, the energy remaining after the test is determined by measuring the maximum deflection of the springs. This means that a weld that exhibits good impact properties will absorb more of the energy of the weight and compress the springs less. Care must be taken to avoid bending of the test specimen as this will result in an error in the test. The cross-tension drop-impact test is used when sheet thicknesses exceed 3.2 mm. Thinner spot welded sheets can be impact tested utilizing a standard shear tension specimen and a pendulum-type Charpy impact machine.

Fig. 9.10 Cross-tension
drop-impact test [*Source* AWS]

Direction of Loading

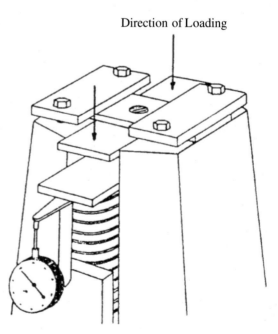

9.3   Nondestructive Evaluation (NDE) and Monitoring Techniques

To reduce the cost of testing, and to avoid destroying large numbers of parts and full car body assemblies, methods of testing other than mechanical testing may be considered. Two general approaches to accomplishing this are NDE of the completed welds, and monitoring of the welds as they are being made.

Two effective NDE techniques for spot welds are radiography and ultrasonic testing, but radiography is very time consuming and expensive so ultrasonic testing is usually the NDE process of choice by the automotive industry. Standard approaches to ultrasonic testing have been used successfully for many years in the evaluation of arc welds. Such techniques have also been used with spot welding, but with limited success.

However, the interest in ultrasonic testing of spot welds has grown significantly recently, primarily due to an improved capability known as phased array ultrasonics. In the phased array approach, the physical probe movement or sweeping required with conventional ultrasonic testing is effectively replaced with "electronic movement." This is achieved through a probe design that contains multiple tiny transducers that can be pulsed individually under the control of computer software (Fig. 9.11). This allows for electronic focusing and steering of the beam. Data from multiple beams can then be assembled to show a cross-sectional image of the weld. Figure 9.12 shows a phased array image above a spot weld cross section. Information from these images can then be compared to defect assessment criteria in applicable codes or specifications to determine whether an acceptable weld has been produced. But it is important to note that mechanical testing continues to be the primary method for evaluating spot weld strength.

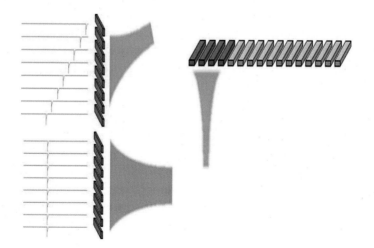

Fig. 9.11 Ultrasonic phased array testing utilizes probes that contain multiple transducers that can be controlled by computer software [*Source* EWI]

Fig. 9.12 Image (top of figure) from ultrasonic phased array testing reveals a "slice" of the spot weld shown at the bottom of the figure [*Source* EWI]

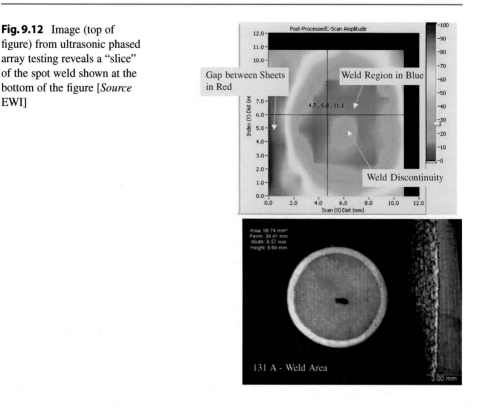

The primary disadvantage of ultrasonic testing is that it is very slow and labor-intensive, and therefore, not practical for testing every spot weld on a vehicle (it is typically only used for sampling and random checks). However, recently a company by the name of Xarion has developed an NDE approach known as laser-excited acoustics (Fig. 9.13) that offers promise for use in high-speed production. As shown in Fig. 9.14, this approach utilizes a pulsed laser on one side of the device and an optical microphone on the other. The laser is pulsed at a certain frequency to create a sound wave of that same frequency which passes through the weld and is then detected by the optical microphone. The detected sound wave is then analyzed to determine the weld size. Because the process is a non-contact process unlike traditional ultrasonic testing which requires a couplant, and it is entirely contained in a small device, it offers significant potential for an automotive assembly line.

Over the last 30–40 years, a significant amount of research has been conducted on a wide range of approaches to in-process monitoring of spot welds. Characteristics of the process that can be monitored and related to weld quality include electrode movement (displacement), sheet temperature, current, voltage, and dynamic resistance, to name a few. However, while much success has been reported in laboratory environments, these approaches generally have not performed so well in production conditions. But more

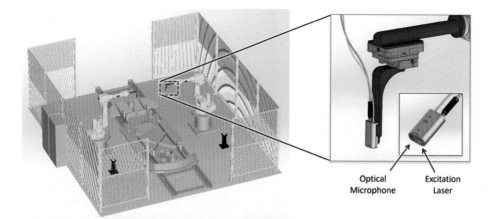

Fig. 9.13 The Xarion laser-excited acoustic system can easily be attached to a robot arm and used for high speed production

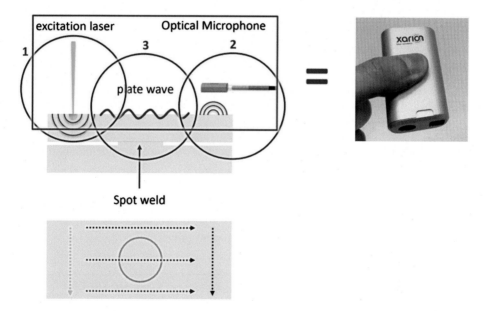

Fig. 9.14 The Xarion laser-excited acoustic method utilizes a single small device that contains both the excitation laser and the optical microphone for detecting the signal

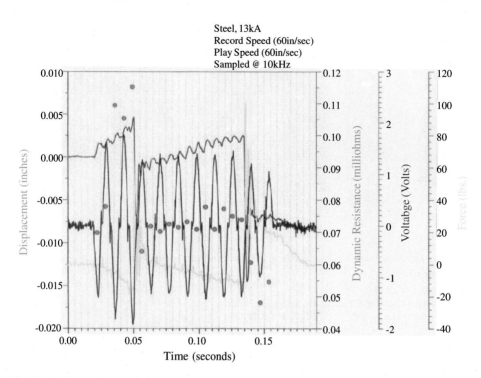

Fig. 9.15 Many characteristics of the spot welding process can be monitored and related to weld quality [*Source* EWI]

recently, as sensor and computer sophistication continue to improve, integration of many spot weld characteristics that can be monitored (Fig. 9.15) and related to weld quality have been incorporated into modern weld controllers. Therefore, monitoring is becoming more common in modern automotive assembly plants. These more advanced controllers can also be used to detect faulty welds and variations in welding conditions and adjust parameters accordingly. This approach is known as adaptive control (Fig. 9.16).

9.4   Computational Modeling of Resistance Spot Welding

The development of computational models (Fig. 9.17) associated with the spot welding process and spot weld testing is a rapidly advancing field. The obvious advantage of computational models for resistance spot welding is the potential for significant reductions in the cost and time associated with process and product development. Models can also provide an improved understanding of spot weld performance (e.g., deformation and failure) when subjected to mechanical loads such as during tensile-shear testing. Modern modeling techniques can provide detailed process simulations to predict temperatures,

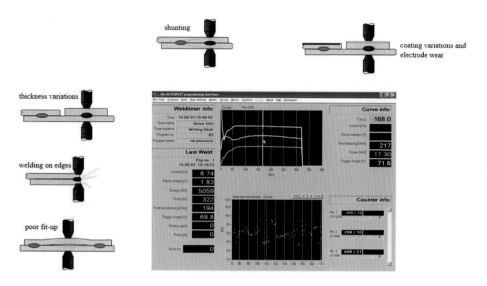

Fig. 9.16   Adaptive control of the spot welding process

microstructures, residual stress/distortion, weld sizes, hardness distributions in the weld and HAZ, and stresses and strains during testing. A variety of commercial modeling software is available which can be customized to consider material type, process parameters, machine characteristics, electrode geometry, etc. Such commercial software ranges from specialized code for spot welding (e.g., Sorpas) to general purpose finite element code such as Abaqus and Ansys.

Figures 9.18 and 9.19 illustrate the ability of models to predict spot weld temperatures. In the first case, the growth of a spot weld is revealed, while the other example shows predicted temperatures on the part being welded. Finally, Figs. 9.20 and 9.21 show the calculated plastic strain distributions during two types of spot weld testing. The model in Fig. 9.20 predicts higher plastic strains concentrated in the HAZ for larger weld sizes, and therefore, a lower probability of an interfacial failure during wedge testing, while the model in Fig. 9.21 shows the strain distribution during a tensile-shear test.

In summary, computational models can be a powerful tool for predicting a wide variety of spot welding conditions and testing outcomes. Once developed, they can be easily modified to reflect changes in material type or thickness, electrode geometries, welding parameters, etc.

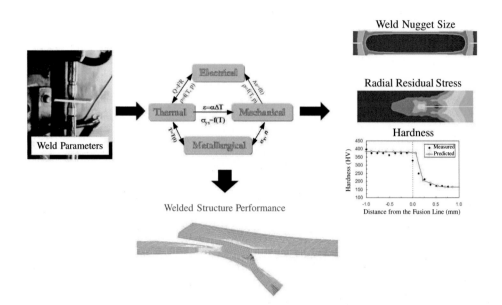

Fig. 9.17 Computational models analyze fundamental spot welding information including electrical, thermal, mechanical, and metallurgical to predict performance

Fig. 9.18 Model predicting
heat distribution during growth
of spot weld [*Source* Simufact]

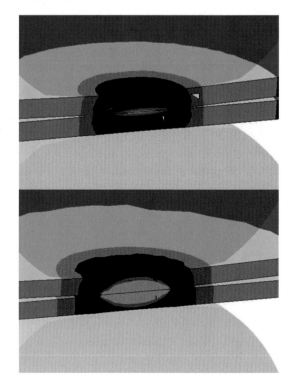

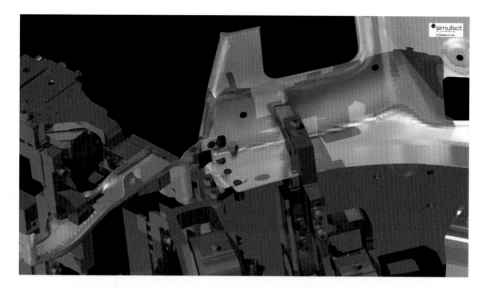

Fig. 9.19 Part temperatures during spot welding [*Source* Simufact]

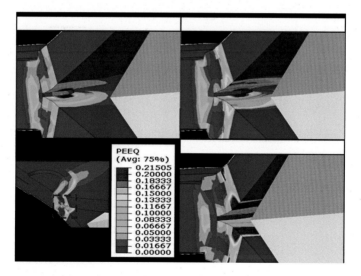

Fig. 9.20 Computational model predicts plastic strain distributions during wedge testing for various sizes of spot welds [*Source* The Ohio State University]

Fig. 9.21 Model of strain
distribution during a spot weld
tensile-shear test

Resistance Spot Welding Production Issues

10

10.1 Electrode Wear

10.1.1 Choice of Electrode Geometry for Long Life

The previous chapters of this book have covered electrode wear as it relates specifically to the materials being welded. More general considerations will be presented here. As discussed previously, spot weld electrode wear is a problem when welding uncoated steels, and becomes a much bigger problem when welding coated steels and aluminum. In most cases, when an electrode wears, its diameter gets larger (Fig. 10.1) which reduces current density, which in turn, reduces the weld nugget size. Since the strength of a spot weld is directly related to its weld size, electrode wear effectively reduces weld strength. As shown in Fig. 10.2, in addition to a decrease in current density, as the electrode face diameter gets larger, the electrode pressure from the applied force gets smaller. As a result, a worn electrode will not only deliver insufficient current density, but insufficient pressure (described on the figure as unit force) as well, eventually resulting in no weld at all. The figure also shows what can happen when choosing the wrong electrode diameter for a given material and thickness. In some cases, however, electrode wear may actually reduce the diameter of the electrode. This can happen if the electrode edges wear more rapidly than the center of the electrode face. The result of this type of unusual wear is excessive current density and expulsion.

Electrode geometry obviously plays a significant role in electrode wear. As discussed previously, there are many geometries to choose from. The choice of geometry may be dictated by the need to keep part marking to a minimum, or other factors such as the need to place a weld close to a flange or in a difficult to reach location. But when electrode wear is the primary consideration, a geometry that provides for the longest life is usually the best choice.

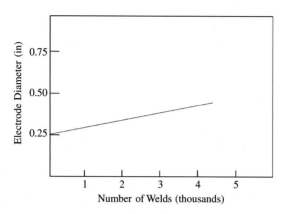

Fig. 10.1 Typical relationship between electrode diameter and number of welds made during electrode life testing

1 Tip Area 400% Too Small	2 Proper Tip Area	3 Tip Area 50% Too Large	4 Tip Area 125% Too Large	5 Tip Area 300% Too Large	6 Tip Area 525% Too Large	7 Tip Area 800% Too Large
0.0123 in.²	0.0491 in.²	0.767 in.²	0.1105 in.²	0.1964 in.²	0.3068 in.²	0.4418 in.²
at	at	at	at	at	at	at
1/8-in. diameter	1/4-in. diameter	5/16-in. diameter	3/8-in. diameter	1/2-in. diameter	5/8-in. diameter	3/4-in. diameter
Unit force 59,600 psi	Unit Force 15,000 psi	Unit Force 9,500 psi	Unit Force 6,500 psi	Unit Force 3,710 psi	Unit Force 2,300 psi	Unit Force 1,650 psi
Current Density 798,000 amps/in.²	Current Density 200,000 amps/in.²	Current Density 127,800 amps/in.²	Current Density 88,000 amps/in.²	Current Density 49,900 amps/in.²	Current Density 31,900 amps/in.²	Current Density 22,200 amps/in.²
RESULT: There is 400% too much force and current. Severe indentation and splitting from high current density.	RESULT: Good spot weld, ideal setup.	RESULT: Only 61% of the required force and current. .Weak spot weld.	RESULT: Only 44% of the required force and current. Stick weld.	RESULT: Only 25% of the required force and current. No weld at all.	RESULT: Only 16% of the required force and current. No weld at all.	RESULT: Only 11% of the required force and current. No weld at all.

Fig. 10.2 Electrode wear results in reduced current density and electrode pressure

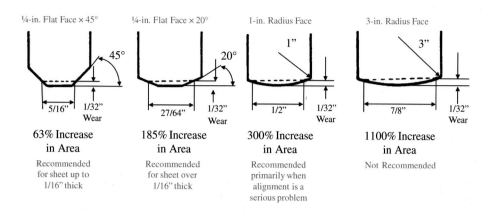

Fig. 10.3 Electrode geometry determines how fast the face diameter changes for a given amount of wear in the vertical direction

However, it may be important to consider that a geometry that provides the longest life might do so at the sacrifice of accelerated electrode wear near the end of its life. This is explained in some detail in Fig. 10.3. Two truncated cone electrode designs are shown on the left side of the figure. The design with the smaller cone angle (20°) might be expected to last longer than the 45° angle because of the greater volume of copper at the end of the electrode, which provides better cooling and therefore maintains strength. While this is true, as indicated in the figure, the disadvantage of smaller cone angles is that for a given amount of electrode, in terms of length of electrode lost (1/32" in the example), there will be a much greater increase in area of the face. Therefore, while the 20° truncated cone electrode design might be expected to last longer before noticeable wear begins, once it begins, wear occurs rapidly and is more likely to produce undersized welds before excessive wear is diagnosed. As shown in the figure, this problem is exacerbated when the design is switched from a truncated design with a flat face to a radius face.

So, when utilizing the common truncated cone shape electrode design, the optimum cone angle is usually a balance between an excessively high angle that may cause the electrodes to get too hot (and therefore, wear rapidly) versus an excessively low angle that will undergo accelerated wear toward the end of its life. Figure 10.4 illustrates this concept of finding a balance and indicates that typical ideal truncated angles are between 30° and 50°, although 45° is most common. Of course, there are many other electrode shapes such as those discussed in Chap. 3, but the concepts associated with electrode wear are the same.

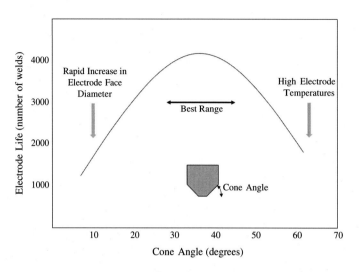

Fig. 10.4 With a truncated electrode, the chosen cone angle should provide for a balance of good cooling but not be susceptible to rapid face diameter changes

10.1.2 Managing Electrode Wear

A common approach to managing electrode wear is to utilize what is known as stepper control. This involves programming the controller to count the number of welds being made and to automatically increase (step up) the current at predetermined numbers of welds. The goal with stepper control is to maintain the current density to maintain the weld size as the electrode wears and the electrode face gets larger. As indicated on Fig. 10.5, an experiment can be conducted which provides information for the stepper control. This experiment involves producing thousands of welds, and generating a current range at selected intervals, such as every 250 welds, as shown in the figure. As the figure shows, electrode wear mandates the need for higher amounts of current to create a minimum weld nugget size. This data can be used to program the stepper control. While this approach is effective for some time, the problem is there is no easy way to make adjustments to increase the force as the electrodes wear. Eventually, the force is insufficient for the size of the electrode face and the amount of current. The result is inconsistent welds and excessive expulsion.

Stepper control may therefore incorporate automatically dressing the electrodes at regular intervals, as well as indicating when the electrodes must be replaced. As before, the counter on the controller counts the number of welds made until it reaches the predetermined number for electrode dressing. In a typical spot welding assembly line, the robot rapidly moves the electrodes to an automatic dressing tool (Fig. 10.6) in a movement that is coordinated with the production flow. When spot welding steels, a more aggressive grinding operation spread out over longer times to maintain the electrode diameter

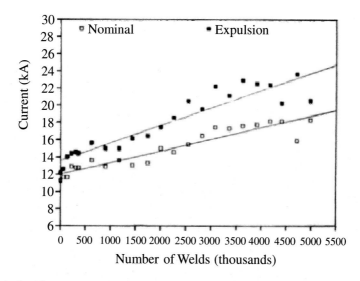

Fig. 10.5 A plot of current range versus number of welds can be used to program the controller

is typically used. But as discussed previously, when welding aluminum, a more effective approach may be to dress more frequently with a less aggressive buffing operation of the electrode face.

Uneven electrode wear sometimes occurs due to a phenomenon known as the "Peltier" effect. This is a type of thermoelectric effect associated with an applied voltage/current at the junction between two dissimilar metals (in this case, the electrodes and sheets being welded). It is only associated with DC current, and results in excessive heating, and therefore electrode wear, on the positive side of the electrodes. This effect can be thought

Fig. 10.6 Automatic spot welding electrode dressing tool

of as opposite to a thermal couple, which produces voltage when exposed to heat. In addition to uneven electrode wear, it may also produce a heat imbalance problem. Some modern equipment can easily address the potential problem of uneven electrode wear by switching the direction of the DC current for each weld.

10.2 Electrode Cooling

Production spot welding systems rely heavily on a continuous flow of water that cools not only the electrodes, but the transformer as well. Many production problems such as rapid electrode wear and transformer failure may arise if the water-cooling system is not well maintained. Electrodes are cooled by the flow of water from a tube positioned in the internal cooling cavity. The water flows out of the tube into the electrode cavity and then exits the cavity out of the back of the electrode. Water cooling effectiveness may be degraded if the tube is not properly positioned or moves during production (Fig. 10.7). If the tube end is too close to the electrode tip (right of the figure), water flow out of the tube may be restricted, whereas if it is too far from the tip (left of the figure), turbulence in the cavity may inhibit good flow out of the electrode.

Other problems common to maintaining a spot welding water-cooling system are corrosion and mineral deposit buildup in the lines. These problems can be controlled by chemically treating the water. Coolers may be used to maintain a cool water temperature, but proper flow of the water is more critical than the temperature. In fact, water temperatures that are too cool should not be used due to condensation concerns. Production spot welding machines utilize water flow gauges that should be monitored to ensure that proper flow rates are being maintained. More detailed information regarding cooling water requirements can be found in the RWMA Resistance Welding Manual.

newpage

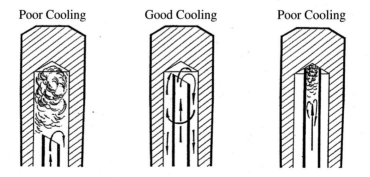

Poor Cooling   Good Cooling   Poor Cooling

Fig. 10.7 The importance of cooling water tube location in the electrode cavity

10.3   **Electrode Alignment and Part Fit-Up**

Established spot welding procedures rely on properly aligned electrodes and good part fit-up. Electrodes that are not aligned will result in expulsion and/or undersized welds. Two common alignment problems are angular and vertical misalignment (Fig. 10.8). Angular misalignment will produce an unequal pressure distribution across the electrode faces while vertical misalignment will reduce the contact area.

Parts that are not properly formed may create a gap in the region where the weld nugget is to be formed (Fig. 10.9). The result is the required electrode force sfor producing a good weld is effectively reduced, potentially resulting in excessive expulsion and an unacceptable weld. Higher weld forces will be required which will increase electrode wear. The problem is magnified with the much higher strength advanced high-strength steels that have greater stiffness. Poor fit-up will effectively mandate that the spot welding machine be used as a press, which of course is not what it is intended for. Therefore, it is very important that the parts are properly formed prior to spot welding.

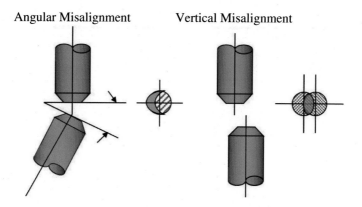

Fig. 10.8 Two common types of spot weld electrode misalignment

Fig. 10.9 Poor part fit-up may
cause spot welding problems

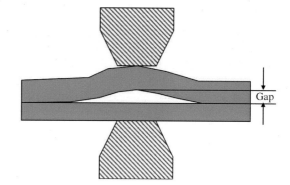

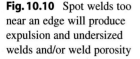 **Fig. 10.10** Spot welds too near an edge will produce expulsion and undersized welds and/or weld porosity

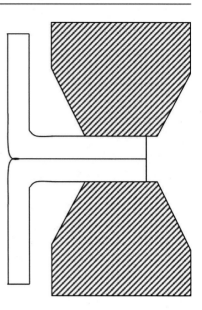

   Parts may be properly fit-up and electrodes properly aligned, but if the spot weld is produced too near an edge (Fig. 10.10), excessive expulsion and an undersized weld and/or weld porosity will be the result. To avoid this problem, Type "D" Offset electrodes that are flat on one side may be used, or the part may need to be redesigned to increase the flange length.

10.4   Pneumatic Versus Electric Servo Guns

In the past, the force supplied by spot welding machines was commonly generated using pneumatic cylinders and various valve systems. But many modern spot welding machines for the automotive industry are switching to electric servomotor driven force systems. These advanced systems offer the advantages of more precise control of electrode force, as well as the elimination of the need for a pressurized air system supply throughout the production facility. They also offer more control when opening the electrodes in between welds, which reduces overall weld time. However, an advantage of pneumatic force is that it offers what is known as good "follow-up." Follow-up is a concept that refers to the machine's ability to maintain the set force during the weld cycle as the electrodes move, such as when there is some indentation. This concept is especially important with projection welding, which involves significant electrode movement as the projections collapse, but can sometimes be important when spot welding as well. One example is when welding aluminum, which due to its high thermal expansion coefficient, undergoes significant expansion and shrinkage of the weld nugget. Poor follow-up may produce excessive expulsion and/or inconsistent welds.

Follow-up is further explained in Figs. 10.11 and 10.12. Good follow-up is achieved when the applied electrode force acts like a spring (Fig. 10.11). If the electrodes move as the weld forms, the spring action will help maintain the set force. In fact, when follow-up is extremely important (such as with projection welding), it is not uncommon for machines to include actual springs added in series to the system. But regardless of whether additional springs are added, pneumatic systems generally supply good follow-up because the cylinder of air that provides the force can be compressed like a spring. And the smaller the cylinder, the better the follow-up.

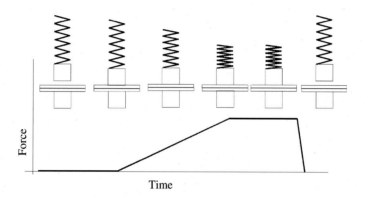

Fig. 10.11   An electrode force system that acts like a spring is needed for good follow-up

Fig. 10.12   Poor follow-up
may result in a drop in force
after current is applied and the
weld forms

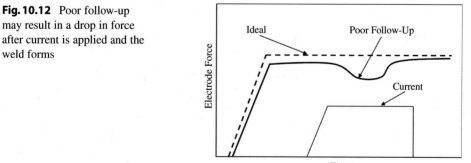

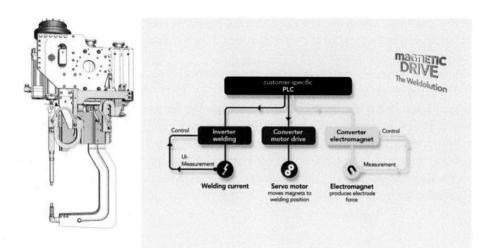

Fig. 10.13 Patented NIMAK magnetic drive system offers improved follow-up in servo-driven force systems [*Source* Otto von Guericke University Magdeburg, Institute of Materials and Joining Technology]

So, whereas traditional servo-driven force systems may offer limited follow-up, some of the most modern systems are becoming much more capable of doing so. For example, a patented system known as "Magnetic Drive" (Fig. 10.13) which combines a more conventional electric servomotor drive for larger strokes with a high dynamic electromagnetic actuator capable of small strokes at high speed, resulting in improved follow-up. This system can also vary and measure force levels during a single weld.

Other manufactures offer modern lightweight low maintenance servo driven systems (Fig. 10.14). Many of these systems have the capability for controlling force and providing force feedback during a single weld. As discussed previously, the ability to control and vary force can be especially important with more challenging spot welds such as multi-sheet stack-ups.

Fig. 10.14 Obara MFDC Spot
Welding System (with Roman
transformer and WTC
controller) with modern servo
gun at the labs at The Ohio
State University

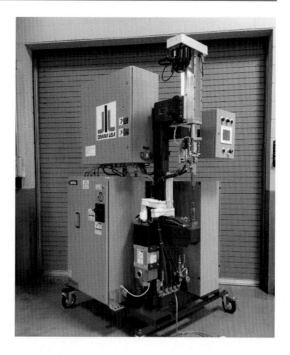

Overview of Resistance Spot Welding Codes and Standards

11

This chapter provides a summary of the various codes and standards pertaining to spot welding that are utilized by automobile manufacturers in the United States. In addition to the codes and standards described below, it is important to note that almost every automotive manufacturer develops there own modified versions of these.

11.1 AWS C1.1

This AWS standard is titled "Recommended Practices for Resistance Welding." This type of standard is not a code or specification in that it doesn't mandate actions but provides advice and recommendations on a wide range of resistance welding process and topics. Included among the spot welding topics covered are suggested welding parameters for common alloy families, including (of course) coated and uncoated carbon and low alloys steels as well as aluminum. There is a section on weld quality, including all the methods for mechanically testing spot welds, ranging from peel testing, to the various forms of tensile testing, to impact and fatigue testing. Other subjects covered include weld bonding, equipment monitoring and maintenance, and safety and health.

11.2 AWS D8.1

The AWS D8.1 specification is titled "Specification for Automotive Weld Quality-Resistance Spot Welding of Steel." It addresses visual acceptance criteria for surface inspection of spot welds, but its major emphasis is on the destructive tests peel, chisel, shear and cross tension. A significant portion of this 28-page specification is dedicated to defining the 8 different fracture modes associated with peel and chisel testing. It also

© The Author(s), under exclusive license to Springer Nature Switzerland AG 2023
M. Kimchi and D. H. Phillips, *Resistance Spot Welding*, Synthesis Lectures on Welding Engineering, https://doi.org/10.1007/978-3-031-25783-4_11

provides minimum acceptable shear and tensile strength values as a function of sheet thickness.

11.3   AWS D8.9

This specification is titled "Recommended Practices for Test Methods for Evaluating the Resistance Spot Welding Behavior of Automotive Sheet Steel Materials." This 65-page comprehensive specification provides much detail regarding standardized test methods for evaluating spot welding behavior of coated and uncoated steels in a laboratory environment. The test methods are not intended to simulate production welding situations, but are to be used to help determine welding currents and ranges, electrode wear, mechanical properties of various weld sizes, effects of paints or lubricants on the sheets, and metallurgical and hardness properties of welds. The testing details addressed include sample preparation, material characterization, welding parameters, test equipment, electrode installation and dressing procedures, endurance (electrode life) testing, weld property testing, and current break-through testing (an approach to determining whether the weld can be made through a lubricant or paint). All yield strengths of steels maybe be tested, but the range of sheet thickness is limited to a minimum of 0.6 mm and a maximum of 3.0 mm.

11.4   ISO Standards

A wide variety of International Organization for Standardization (ISO) standards pertaining to spot welding are available to the automotive manufacturing industry. It is beyond the scope of this book to summarize all of them, but the topics will be mentioned here. Each of the following topics represents a specific ISO standard that can be purchased directly from ISO, or from organizations such as AWS:

- "Specification and Qualification of Welding Procedures for Metallic Materials—Welding Procedure Test—Spot, Seam and Projection Welding";
- "Resistance Welding Equipment—Transformers—General Specifications Applicable to All Transformers";
- "Mechanical Joining—Destructive Testing of Joints—Specimen Dimensions and Test Procedure for Tensile Shear Testing of Single Joints";
- "Quality Requirements for Welding—Resistance Welding of Metallic Materials—Part 1: Comprehensive Quality Requirement";
- "Quality Requirements for Welding—Resistance Welding of Metallic Materials—Part 2: Elementary Quality Requirements";

- "Resistance Welding—Procedures for Determining the Weldability Lobe for Resistance Spot, Projection and Seam Welding";
- "Electrode Taper Fits for Spot Welding Equipment—Dimensions".

11.5 RWMA

The Resistance Welding Manufacturers Association (RWMA) "Resistance Welding Manual" is a well-known book that covers all topics related to all resistance welding processes. In addition, RWMA provides bulletins that cover topics such as electrodes, machine installation, maintenance, and operation, and standards pertaining to the mechanical and electrical aspects of a resistance welding machine.

Notes: Previous spot welding standards AWS D8.6 and D8.7 have been discontinued. Also, the standards discussed above are most applicable to U.S.-based automotive companies. Standards that may be used by automotive manufacturers based in other countries (i.e., some Japanese automotive companies utilize Japanese Industry Standards-JIS) are not addressed in this book.

Printed in the United States
by Baker & Taylor Publisher Services